Sustainable Aviation

Series Editors

T. Hikmet Karakoc, Eskisehir Technical University
Eskisehir, Türkiye

C. Ozgur Colpan, Department of Mechanical Engineering
Dokuz Eylül University
Buca, Izmir, Türkiye

Alper Dalkiran, School of Aviation
Süleyman Demirel University
Isparta, Türkiye

The Sustainable Aviation book series focuses on sustainability in aviation, considering all aspects of the field. The books are developed in partnership with the International Sustainable Aviation Research Society (SARES). They include contributed volumes comprising select contributions to international symposiums and conferences, monographs, and professional books focused on all aspects of sustainable aviation. The series aims at publishing state-of-the-art research and development in areas including, but not limited to:

- Green and renewable energy resources and aviation technologies
- Aircraft engine, control systems, production, storage, efficiency, and planning
- Exploring the potential of integrating renewables within airports
- Sustainable infrastructure development under a changing climate
- Training and awareness facilities with aviation sector and social levels
- Teaching and professional development in renewable energy technologies and sustainability

T. Hikmet Karakoc • Can Ozgur Colpan
Alper Dalkiran
Editors

Sustainable Materials and Manufacturing Techniques in Aviation

Springer

SARES
INTERNATIONAL SUSTAINABLE AVIATION
AND ENERGY RESEARCH SOCIETY

Editors
T. Hikmet Karakoc
Faculty of Aeronautics and Astronautics
Eskisehir Technical University
Eskisehir, Eskisehir, Türkiye

Information Technology Research and Application Center
Istanbul Ticaret University
Istanbul, Türkiye

Alper Dalkiran
School of Aviation
Süleyman Demirel University
Keciborlu, Isparta, Türkiye

Can Ozgur Colpan
Makina Mühendisliği Bölümü
Dokuz Eylül Üniversitesi
Izmir, Izmir, Türkiye

ISSN 2730-7778 ISSN 2730-7786 (electronic)
Sustainable Aviation
ISBN 978-3-031-62986-0 ISBN 978-3-031-62987-7 (eBook)
https://doi.org/10.1007/978-3-031-62987-7

This Springer imprint is published by the registered company Springer Nature Switzerland AG
The registered company address is: Gewerbestrasse 11, 6330 Cham, Switzerland

Preface

In an era defined by environmental consciousness and technological innovation, the aviation industry finds itself at a pivotal juncture. The escalating demand for air travel, coupled with the imperative to mitigate carbon emissions and enhance sustainability, has spurred a wave of transformative advancements. This edited volume serves as a comprehensive exploration into sustainable materials and manufacturing techniques in aviation, offering insights into the multifaceted challenges and promising solutions shaping the future of flight.

This edited book includes five chapters. Chapter 1 provides a comprehensive analysis of the aviation industry's multifaceted strategy aimed at bolstering efficiency and mitigating its carbon footprint. Through an exploration of advanced materials such as carbon fiber-reinforced polymers (CFRP) and graphene, alongside the integration of additive manufacturing techniques, the pursuit of sustainability unfolds. This rigorous examination extends to the meticulous assessment of pivotal ratios, discerning their influence on both energy consumption and the paramount concern of flight safety. Chapter 2 delves into the pressing challenge of climate change and the industry's quest for sustainable composite materials. As the aviation landscape evolves, the exploration of biocomposite materials and their potential applications in aircraft design emerges as a critical focal point, underscoring the need for collaboration and compliance with regulatory standards. In Chap. 3, the spotlight shifts to additive manufacturing technology as a catalyst for reimagining aerospace structures. By harnessing the capabilities of modern additive manufacturing techniques, the industry stands poised to revolutionize production processes, optimize structures, and pave the way for a more sustainable future in aerospace operations. Chapter 4 offers a glimpse into the realm of artificial intelligence (AI) and its transformative impact on aircraft production systems. Through insightful case studies, readers gain an appreciation for the integration of AI technologies in streamlining manufacturing processes, optimizing production schedules, and laying the groundwork for advanced digital smart factories. Finally, Chap. 5 explores the role of digital twin architecture in enhancing aircraft sustainability. By leveraging real-time data analysis and predictive capabilities, the proposed architecture holds the promise of optimizing engine performance, reducing fuel consumption, and mitigating carbon emissions—a testament to the power of technological innovation in fostering sustainable aviation practices.

As editors of this volume, we are honored to present a collection of chapters that not only shed light on the challenges facing the aviation industry but also offer innovative solutions and pathways toward a more sustainable future. It is our hope that this compilation will inspire collaboration, spark dialogue, and inform future endeavors aimed at redefining the trajectory of aviation in the twenty-first century.

Istanbul, Türkiye — T. Hikmet Karakoc
Izmir, Türkiye — Can Ozgur Colpan
Keciborlu, Türkiye — Alper Dalkiran

Contents

Energy Savings from New Materials and Processes in Aviation

Tamer Saraçyakupoğlu

Nomenclature

AOC	Air Operator Certificate
ATF	Aviation Turbine Fuel
AM	Additive Manufacturing
BALD	Bleed Air Leak Detector
B/F	Buy-To-Fly
CAD	Computer-Aided Design
CFD	Computational Fluid Dynamics
CO	Carbon Monoxide
CFRP	Carbon Fiber Reinforced Polymers
DED	Directed Energy Deposition
DMD	Direct Metal Deposition
DMLS	Direct Metal Laser Sintering
DOC	Direct Operating Cost
EBM	Electron Beam Melting
EBFF	Electron Beam Freeform Fabrication
ECS	Environmental Control System
ELSP	Enhanced Lightning Strike Protection
FEA	Finite Element Analysis
FDM	Fused Deposition Modeling
HC	Hydrocarbons
HCG	Hydrophobic Conductive Graphene
HCL	High Cyclic Load

T. Saraçyakupoğlu (✉)
Turkish Aerospace Industries, Ankara, Turkey
e-mail: dr.tamer@tamersaracyakupoglu.com.tr

T. H. Karakoc et al. (eds.), *Sustainable Materials and Manufacturing Techniques in Aviation*, Sustainable Aviation, https://doi.org/10.1007/978-3-031-62987-7_1

HIP	Hot Isostatic Press
HPC	High-Pressure Compressor
HPT	High-Pressure Turbine
ICAO	International Civil Aviation Organization
IPC	Intermediate Pressure Compressor
ISA	International Standard Atmosphere
LCL	Low Cyclic Loads
LFF	Laser Freeform Fabrication
LIG	Laser Induced Graphene
LMD	Laser Metal Deposition
LOM	Laminated Object Manufacturing
LENS	Laser Engineered Net Shaping
LPC	Low-Pressure Compressor
LPT	Low-Pressure Turbine
MRO	Maintenance, Repair, and Overhaul
MSL	Mean Sea Level
MTOW	Maximum Takeoff Weight
NDI	Non-Destructive Inspection
OAP	Outside Ambient Pressure
OAT	Outside Ambient Temperature
PBF	Powder Bed Fusion
PM	Particulate Matter
RCL	Random Cyclic Loads
SLM	Selective Laser Melting
SLS	Selective Laser Sintering
SMD	Shaped Metal Deposition
S/W	Strength-to-Weight
T/W	Thrust-to-Weight
UAM	Ultrasonic Additive Manufacturing
UAV	Unmanned Aerial Vehicle
UN	United Nations
UTS	Ultimate Tensile Strength
WAAM	Wire and Arc Additive Manufacturing

Subscripts

CO_2	Carbon Dioxide
H_2O	Water Vapor
NO_X	Nitrogen Oxides
SO_2	Sulphur Dioxide
SO_X	Sulphur Oxides

1 Introduction

This chapter will provide information regarding energy savings that can be gained using new materials and processes. In addition, there will be information about aircraft manufacturing techniques, maintenance, and aviation-grade materials.

Air vehicles are designed and manufactured to operate under extraordinary and challenging conditions. According to International Standard Atmosphere (ISA) conditions, the pressure is 1 Atm (1013.25 hPa) and the temperature is 15 °C air temperature at Mean Sea Level (MSL). However, the Outside Ambient Temperature (OAT) is −55 °C −60 °C, and the Outside Ambient Pressure (OAP) is ¼ at the cruise flight level (approximately 33,000 feet to 36,000 feet) of a commercial passenger aircraft, as shown in Fig. 1. Besides, the amount of oxygen in this environment is one-third of MSL. If a person were to emerge into this harsh environment, who would freeze within approximately 20 seconds, whose blood would begin to boil because of low pressure, and who would suffocate due to lack of oxygen. With this regard, one can easily claim that the air vehicles are designed using state-of-the-art technologies which can endure these severe conditions.

For example, an Airbus A380, which has a Maximum Takeoff Weight (MTOW) of approximately 560 tons and accommodates 853 passengers, can fly safely even in winds that sometimes reach 300 mph. Besides, its speed can increase to about 900 km/h while the passengers enjoy a comfortable and smooth journey.

The companies manufacturing these state-of-the-art air vehicles must have certified facilities, technological equipment, and well-educated staff to cope with these challenging conditions. These complex and high-level qualities necessitate total technical growth, so the aviation industry often represents a country's technological level [1–3].

As we look at the background of these cutting-edge technologies, aviation industry has been in constant development since the Wright brothers who achieved the first controllable powered flight on December 17, 1903. The developments primarily consider flight safety, which can be handled with robust materials and

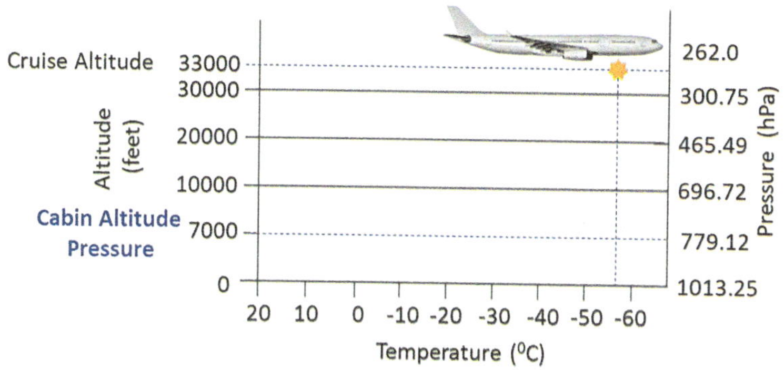

Fig. 1 Outside environment conditions of a commercial passenger cruise altitude

onboard-proof technologies. On the other hand, while engineering studies have been carried out to increase flight safety, the necessity of producing more cost-effective parts has emerged to overcome the increasingly fierce and well-known competition in the aviation industry.

Airbus and Boeing are known to dominate commercial passenger aircraft, as shown in Fig. 2. The closest follower is Embraer. It can be seen that there is a big gap between the top two manufacturers and the third one.

Notably, orders dropped significantly in 2019, 2020, and 2021 because of the global pandemic. In addition, particularly for Boeing, Boeing 737Max accidents augmented the drop in aircraft deliveries because of vast cancellations.

The International Civil Aviation Organization (ICAO) is the top organization of the United Nations (UN) in the context of civil aviation regulations. ICAO mainly regulates the civil aviation industry. ICAO declares that the number of passengers traveling by air worldwide has increased twofold every 15 years since 1977. It is expected to have a similar trend in the future [7]. Based on current projections, air travel is expected to grow annually between 3% and 5% [8].

It means that there will be increase in orders synchronizing with transportation demand in the future. This market widening is not only for aircraft manufacturing but also for Maintenance, Repair, and Overhaul (MRO) activities. As seen in Fig. 3, MRO activities are generally 3–4 times bigger than procurement over the lifespan of an aircraft.

It should be underlined that the stated market enlargement brings some handicaps in terms of environmental concerns. The aviation industry's current carbon dioxide emissions comprise approximately 2–4% of total emissions [10, 11].

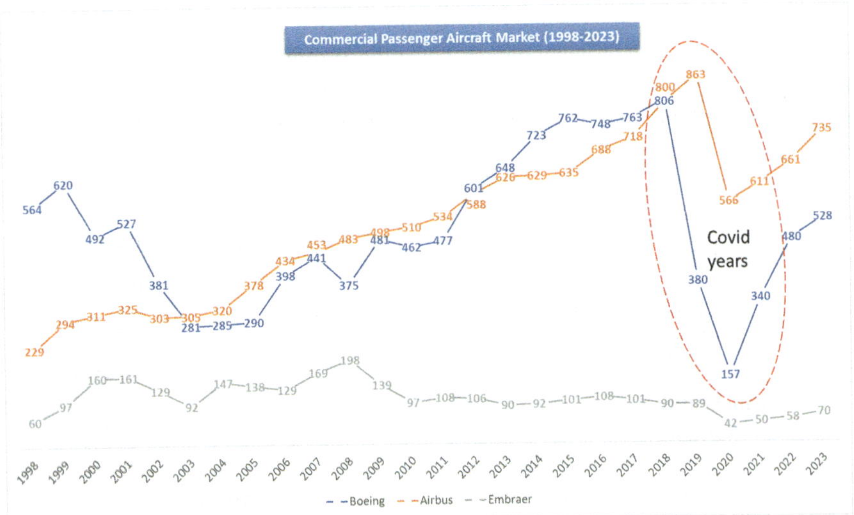

Fig. 2 Number of gross orders for Airbus, Boeing, and Embraer Commercial Passenger Aircraft between 1998 and 2023 [4–6]

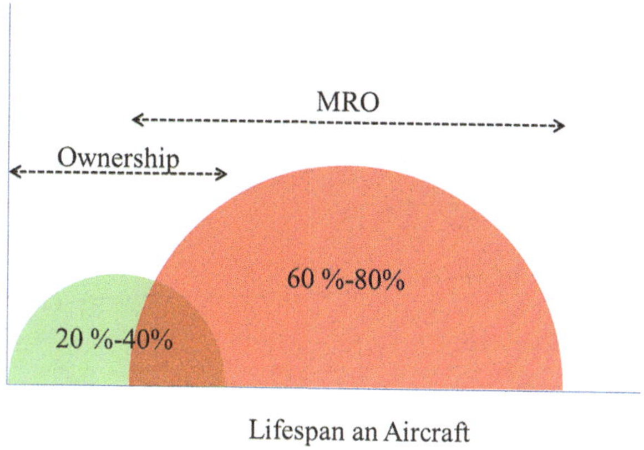

Fig. 3 Ownership and MRO cost ratio within the life cycle [9]

Another crucial issue is that while performing energy-saving activities, the researchers are assisting in decreasing the carbon footprint.

It is important to understand the concept of Direct Operating Cost (DOC), which reflects an aircraft's operational expenses. This concept is closely related to energy efficiency and has an impact on its carbon footprint.

1.1 Direct Operating Cost (DOC)

For saving energy, the term DOC is of crucial importance. The term DOC describes the airline's total costs when running the whole fleet. It covers all expenses directly linked with an airliner's daily operations, including fuel, crew salary, maintenance, insurance, and aircraft leasing or amortization. As mentioned before, there's fierce competition in the aviation industry. Therefore, the airlines need to calculate DOC because it provides an unambiguous overview of their financial performance relative to operational efficiency. As it enables airlines to determine the most economical methods of fleet operation, it plays a crucial role in strategic decision-making. Figure 4 displays the overall airline DOC distribution and breakdown of the items.

Figure 4 shows that 33.4% of fuel has the highest input for airlines. With a 10.6%, aircraft ownership costs significantly lower than fuel. The remaining cost items decrease gradually with very minor variations. Undoubtedly, airlines and aircraft production businesses value any effort to reduce fuel consumption and, consequently, DOC.

So, how can reducing fuel consumption, meaning saving energy in flight operations, be achieved? The answer to every engineering question contains an optimization. With this regard, the answer is to produce aircraft parts that are more innovative, lighter, more durable, and have higher mechanical performance, as presented in Fig. 5. Kaufmann et al. [13] showed that there is a direct correlation between part weight, cost, and mechanical performance.

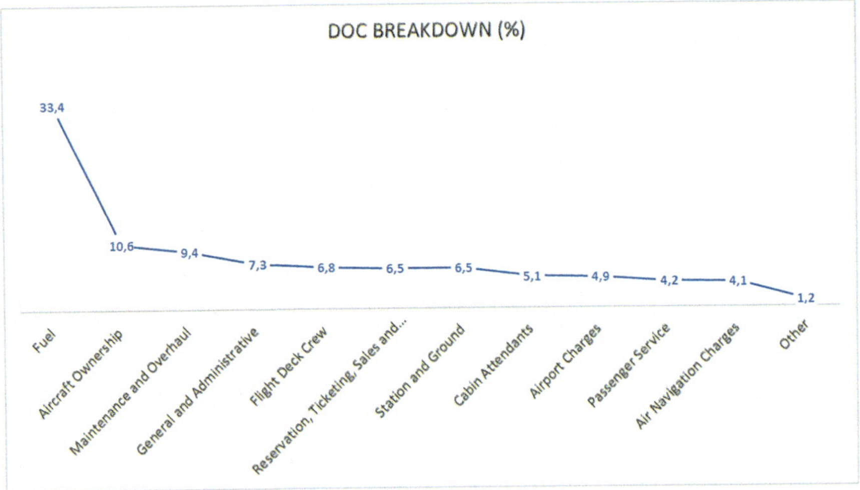

Fig. 4 DOC breakdown of a commercial passenger aircraft operation breakdown. (Derivated from the information based on [12])

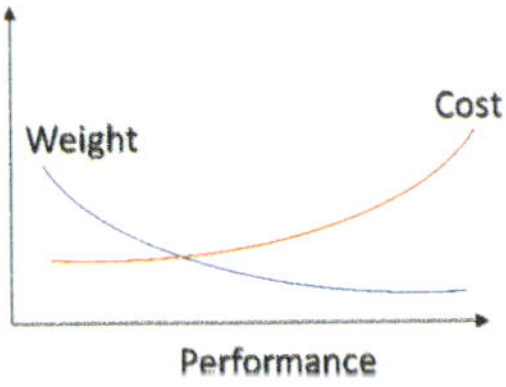

Fig. 5 Optimization amongst weight, cost, and performance for airworthy parts

As seen in Fig. 5, the weight and cost have to be traded off depending on how well the structural component performs. In general, higher performance causes higher production costs. In other words, trying to reduce the weight of an aircraft part does not only focus on maintaining the mechanical properties of the part. At the same time, the part's production cost must also be considered. Every engineering study is an optimization exercise between cost, relevant requirements, and customer demands. In other words, optimizations should be made without compromising flight safety to ensure that production and maintenance costs are bearable. At the same time, energy saving can be achieved by reducing part weight. This is where the approach of using lighter materials with the same strength or acceptable mechanical properties comes into play. When we look at the history of aviation, it is observed that the aircraft production process, which started with wood and steel, has evolved towards using aluminum, titanium, and Carbon Fiber Reinforced Polymers (CFRP) materials due to the need for lighter materials. Figure 6 shows a ratio of the densities of the steel, titanium alloys, 7075 aluminum alloy, magnesium, and CFRP, which are commonly used on aircraft.

The density of steel is about 7.7–8.0 gr/cm^3, and the density of aluminum 7075 alloy is 2.7 gr/cm^3, while CFRP's density is approximately 1.6 g/cm^3 [14]. In other

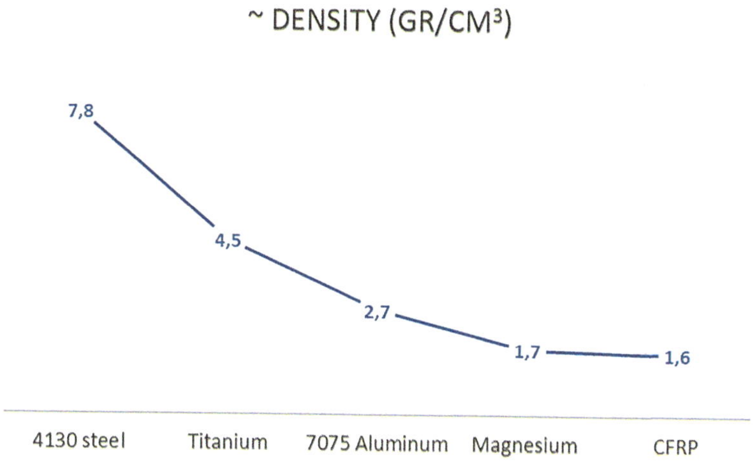

Fig. 6 The density comparison of the widely used aviation materials

words, steel is five times heavier than CFRP. Even this comparison explains why CFRP usage is widespread in the aviation industry. Moreover, composite materials provide many advantages, including a high strength-to-weight ratio, better corrosion resistance, enhanced fatigue resistance, and low maintenance costs [15]. The use of composite materials such as CFRP has a significant impact on lowering CO_2 emissions.

The amount of fuel burned is critical to the efficiency of aircraft operation. According to empirical data, every kg of additional weight requires an extra 0.02 to 0.03 kilograms of gasoline per 1000 kilometers. This statement is called the Marginal Fuel Burn Rate (MFB). Commercial passenger aircrafts generally use Aviation Turbine Fuel (ATF) as their primary fuel source. The prices of ATF can vary depending on the tax tariffs applied by different countries and, in some cases, even from state to state and city to city. As of March 2024, the average price of 1 liter of ATF is around 0.64 USD. For every additional 1 kg of cargo taken on board during a 1000 km flight, there is a cost increase of approximately 0.012 USD. That is one of the reasons that the airlines put rules for weight on luggage and cabin-size bags. The contrary is also true: a 1-kg weight reduction saves 0.02 to 0.03 kg of gasoline every 1000 kilometers [16]. Using less fuel per kilometer should not only be considered energy savings. Utilizing less fuel also reduces the amount of gas emissions and other aerosols emitted by airplanes, as shown in Fig. 7.

During regular operation, fuel burning produces gaseous particles in the exhaust of engined aircraft. As depicted in Fig. 8, these pollutant emissions are:

- Nitrogen oxides (NO_X).
- Unburned hydrocarbons (HC).
- Carbon monoxide (CO), sulfur oxides (SO_X).
- Carbon dioxide (CO_2).
- Water vapor (H_2O).
- Particulate matter.
- Soot.

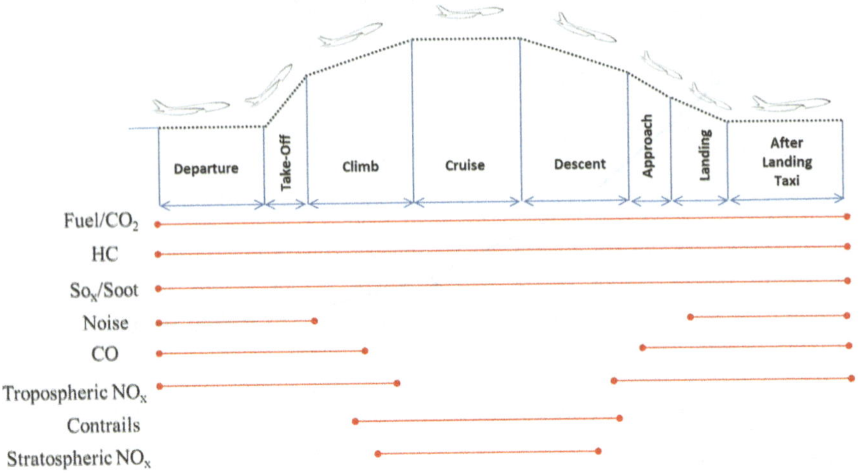

Fig. 7 The environmental implications of aircraft operation per flight phases (The figure was re-illustrated based on [17])

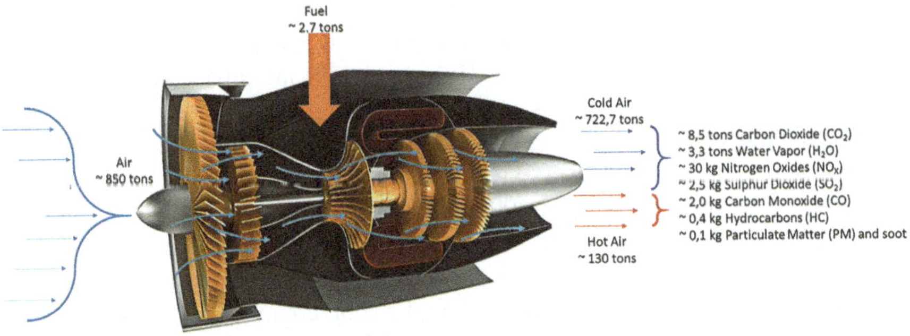

Fig. 8 Emissions and pollutants from a typical single-aisle twin-engine commercial passenger aircraft during a 1-hour flight. (Re-illustrated based on [18])

The exhaust contains approximately 70% carbon dioxide and about 30% water vapor, which are undesirable emissions but are unfortunately unavoidable because fossil fuels are utilized for propulsion [10].

Reducing 1 kg from Boeing 747–400 (MTOW is 396,890 kg) can lower carbon emissions to 0.94 kg. Similarly, reducing 1 kg from Airbus A330–300 (MTOW is 242,000 kg) can decrease carbon emissions up to 0.475 kg [19]. According to a specific study by Boeing, removing 0.435 kg (1 pound) of weight from each aircraft from the American Airlines fleet could save approximately 26.190 kg (11,000 gallons) of gasoline per year [20].

Finally, the fierce competition in the aviation market forces aircraft manufacturer companies to produce lighter aircraft, maintaining mechanical performance and at least the same payload capacity.

1.2 Three Critical Ratios for Determining the Energy Saving in the Aviation Industry

As depicted in Fig. 9, there are three important ratios for determining energy savings in the aviation industry. These are "thrust-to-weight ratio," "buy-to-weight ratio," and "strength-to-weight ratio."

As stated previously, the aviation sector has always been at the cutting edge of innovation and growth, constantly searching for methods to increase efficiency, reduce expenses, and consequently save energy. As fossil fuels become limited and environmental concerns increase, the sector has prioritized developing ways to conserve energy and lower emissions. In this regard, the "thrust-to-weight ratio," "buy-to-weight ratio," and "strength-to-weight ratio" became three critical ratios commonly used to assess and define an aircraft's energy-saving potential.

The thrust-to-weight ratio is a key factor in determining an aircraft's power and efficiency. This ratio is calculated by comparing the amount of thrust the aircraft generates to its total weight. A high thrust-to-weight ratio results in improved maneuverability, fuel efficiency, and the ability to accelerate and climb quickly. Conversely, a lower ratio can lead to sluggish performance and longer takeoff distances.

Another essential ratio to consider when calculating energy-saving potential is the buy-to-weight ratio, which assesses the aircraft's cost compared to MTOW. This ratio considers materials used in construction, aerodynamic design, and engine efficiency. A lower buy-to-weight ratio indicates that an aircraft is more fuel-efficient and cost-effective, making it a more desirable choice for airlines aiming at reducing energy usage.

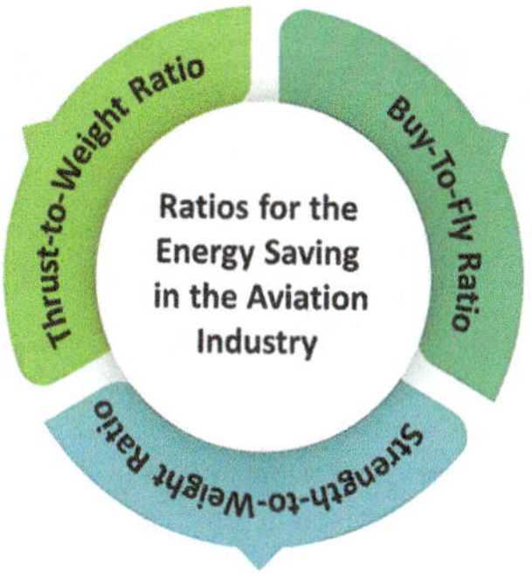

Fig. 9 Three important ratios for determining the energy saving in the aviation industry

Finally, the strength-to-weight ratio, which evaluates an aircraft's structural strength relative to its weight, is crucial for safety and fuel efficiency. A durable and lightweight aircraft can carry more oversized cargo with less fuel, optimizing its energy efficiency. Furthermore, a better strength-to-weight ratio reduces an aircraft's overall maintenance and repair expenses, resulting in additional savings for Air Operator Certificate (AOC) holder companies and/or airlines.

Conclusionally, these three ratios are critical criteria for the aviation sector's efforts to achieve sustainability and cost-effectiveness.

1.2.1 Thrust-to-Weight (T/W) Ratio

The thrust-to-weight ratio, often known as the "T/W ratio," defines the amount of thrust a jet engine produces to the aircraft's MTOW. A high thrust-to-weight ratio indicates that the engine can propel the aircraft with less fuel, resulting in significant energy savings. This ratio is frequently a critical factor in the design and development of new aircraft and in choosing engines for existing ones.

The thrust-to-weight ratio is a crucial term in aviation science since it determines the overall performance and efficiency of the aircraft. It is defined as the "maximum amount of thrust an aircraft engine can produce concerning the aircraft's weight" and computed using the formula "T/W ratio = Maximum Thrust (in S.I. units—in newtons)/Maximum Takeoff Weight (in newtons)." The weight (N) is calculated by multiplying the mass (kg) by the gravity acceleration (m/s^2). It should be noted that thrust can also be measured in pound-force (lbf) if the weight is recorded in pounds (lb). It is noteworthy that thrust must be measured under controlled conditions (Engine cell rooms) to compare the initial T/W ratio.

It can be seen that inefficient flight operations were carried out due to the low power generated by the engines used in the aircraft built in the early years of aviation. In other words, there is a linear correlation between the durability of the airframe and the power produced by the aircraft engine. The fuselages of airplanes with powerful engines must also have high mechanical performance to meet this power. In this concept, two iconic aircraft require mentioning in military aviation history. The first jet-powered Heinkel He-178 (Fig. 10a) and Messerschmitt Me-262

Fig. 10 (**a**) First turbojet powered single engine aircraft Heinkel He-178 Aircraft, (**b**) First turbojet powered twin engine aircraft Messerschmitt Me-262. (Images courtesy of the correspondent author)

(Fig. 10b) aircraft were being operated under harsh conditions of the Second World War. They were both manufactured using steel and strong rivets rather than wood or fabric.

After many attempts, the Messerschmitt Me-262 became the first mass-produced turbojet-powered fighter aircraft. When both aircraft are compared, it can be seen that the previously produced He-178 was single-engine, and the later produced Me-262 was twin engine. The stated aircraft that produces higher thrust is needed over time, depending on war conditions. In this regard, the MTOW of the He-178 aircraft was 1998 kg, and the HeS 3B engine used in this aircraft has 1102 lbf. The MTOW of the Me-262 aircraft was 7130 kg, and the Junkers Jumo 004 engine used in it had 1848 lbf.

The thrust-to-weight ratio is essential for the engineers to determine if the aircraft will be single-engine or twin-engine. Moreover, the engine configuration, operation altitude, desired climb rate, and weather conditions are all issues that might influence an aircraft's T/W ratio. Higher-thrust engines are installed in military and high-performance aircraft, resulting in a higher T/W ratio. Flying at higher altitudes, where the air density is low, can also diminish the T/W ratio due to the smaller amount of oxygen available for combustion.

A high T/W ratio, as indicated in Table 1, demonstrates a strong aircraft with great acceleration, climb rate, and maneuverability.

Unsurprisingly, military fighter aircraft have a higher T/W ratio than commercial passenger aircraft. While jet engines, primarily used for military purposes, are expected to provide sudden acceleration and endurance under challenging maneuvers, engines used in commercial airliners are needed for energy saving.

Also, the T/W ratio is used by engine manufacturers to define the ratio of the energy produced by the engine per the engine's own weight. It is essential to

Table 1 Aircraft types and respected thrust, MTOW, and T/W Ratio

Aircraft	Type	Number of the engine	Maximum thrust (kN)	MTOW (Kg)	Thrust-to-weight
F22 raptor	Jet fighter	Twin	2 x 155.69	38.000	0.835292
Sukhoi-Su-35	Jet fighter	Twin	2 x 141.9	35.000	0.826562
F16 fighting falcon	Jet fighter	Single	129.00	21.722	0.603980
F-35 lightning II	Jet fighter	Single	177.93	31.751	0.571237
Boeing 747-8	Wide body	Four	4 x 295.80	447.696	0.269405
Boeing 767-400 ER	Wide body	Twin	2 x 269.60	204.116	0.269280
Airbus A340	Wide body	Four	4 x 145.00	275.000	0.214994
Airbus A300-600	Wide body	Twin	2 x 273.60	170.500	0.327154
Boeing 707	Narrow body	Four	4 x 84.40	151.315	0.227432
Boeing 727-200	Narrow body	Three	3 x 71.20	95.100	0.228956

Re-arranged based on [21]

distinguish the T/W ratio used for aircraft from the T/W ratio used for engines. Since the engine weighs less than the aircraft on which it is mounted, the T/W ratio used for the engine will be larger than that employed for the airplane. For example, the F110-GE-129 engine in the F16 aircraft has a T/W ratio of 4.38:1 at intermediate power and 7.50:1 when the afterburner is activated. However, as given in Table 1 (third row), the T/W ratio of the F16 aircraft is 0.603980. In the full-power variant, the Merlin 1D rocket engine has a thrust-to-weight ratio of 180.1, nearly twice that of the Saturn V first-stage rocket F-1 engine (94.1) [22].

It is noteworthy that increasing the thrust of an engine usually means adding more components to the engine, such as additional fuel pumps and heat exchangers, which results in an increase in weight. This can ultimately lead to a lower T/W ratio and reduce the engine's overall efficiency.

1.2.2 Buy-To-Fly (B/F) Ratio

In the aviation industry, the term buy-to-fly ratio must be considered when discussing energy saving. The buy-to-fly ratio is the weight ratio between a manufactured end part and its original raw material [23]. In the aviation industry, the B/F ratio varies from 10:1 to 15–20:1 and can reach 40:1 for highly sophisticated components [24]. Some additively manufactured aviation-grade parts have a buy-to-fly ratio as low as 1:1. For example, Joint Strike Fighter (F-35) airplanes Bleed Air Leak Detector (BALD) brackets shown in Fig. 11 were EBM additively manufactured with a buy-to-fly ratio close to 1:1, resulting in a 50% cost savings for the titanium alloy [25] as opposed to the 20:1 using traditional 5-axis milling processes [26].

It's known that the raw materials needed in aviation and the aircraft parts made from these resources are not cheap. They are costly. As a result, it is critical to produce as little waste as possible when transitioning from raw material to ready-to-takeoff parts. For traditional methods, compared to additive manufacturing, the

Fig. 11 BALD bracket manufactured mounted on joint strike fighter

buy-to-fly ratio for airworthy parts is high. Table 2 shows the weight of workpieces (billets), the weight of final parts, and the buy-to-fly ratios.

As can be seen, using legacy while ready-to-take-off part manufacturing, in some cases, causes massive wastage. For example, 657 kg of the aluminum workpiece is machined for the wing rib, and only 18 kg of the final part is gained. The buy-to-fly ratio for the wing rib is 37:1. Meaningly, 97% of the billet became wastage even if the scraps are not calculated. Undoubtedly, this high buy-to-fly ratio directly increases the cost of the ready-to-use final part.

1.2.3 Strength-to-Weight (S/W) Ratio

The strength-to-weight ratio quantifies a material's strength over to its weight. The aviation industry has attempted to improve this ratio by adding composite materials to the primary aircraft structure. Composite materials provide a higher strength-to-weight ratio than standard metals, resulting in lower structural mass and increased fuel economy, as shown in Fig. 12.

Sinha et al. [28] investigate the application of composite optimization to lower the weight of a high aspect ratio composite wing aircraft model. The study found that the composite wing has a higher strength-to-weight ratio than an aluminum wing.

The strength-to-weight ratio of an aircraft relates to its ability to carry a specific amount of load or withstand forces while being lightweight. This ratio is essential for a variety of reasons. Firstly, a better strength-to-weight ratio enables an airplane to carry heavier cargo while remaining within overall weight constraints. The strength-to-weight ratio allows airlines to transport more payloads (people and/or cargo), enhancing profitability and efficiency. Secondly, a high strength-to-weight ratio improves fuel efficiency.

By decreasing the weight of the aircraft structure, fuel consumption can be significantly lowered, resulting in financial savings and a lower environmental effect. Third, a robust, lightweight aircraft structure improves safety. It diminishes the likelihood of structural failure while increasing the aircraft's overall resilience during flight [29]. Furthermore, using composite materials in aviation has significantly

Table 2 The weight of workpieces (billets), the weight of final parts, the buy-to-fly ratios

Component/end-part	Weight of the billet (kg)	Weight of end product (kg)	BTF ratio
Intercase	182	30	6:1
Simple duct flange 1	67	11.1	6:1
Simple duct flange 2	67	7.7	8.7:1
Complex duct flange 1	149	7.7	19.4:1
Complex duct flange 2	207	10.3	20:1
Large blisk	810	97	8.4:1
Wing rib	657	18	37:1

Re-arranged based on [27]

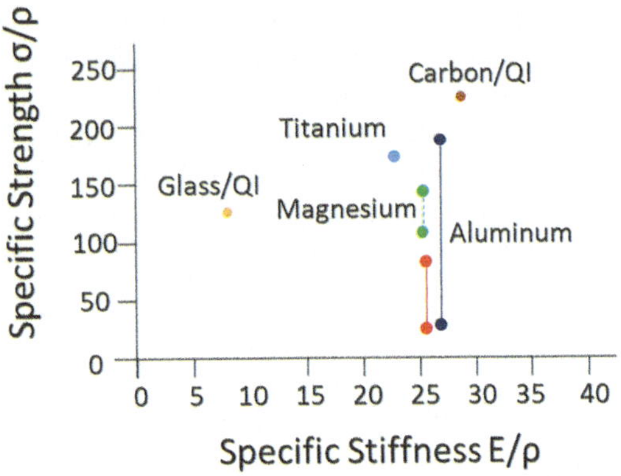

Fig. 12 Specific strength and stiffness of widely used aviation materials. (Re-illustrated based on [13])

improved aircraft strength-to-weight ratio. CFRP has significantly better specific strength compared to aluminum or titanium, yet their specific moduli are similar. A composite component built for rigidity will have a better safety factor against material failure than a metallic component.

A high strength-to-weight ratio allows for composites' good fatigue performance, which reduces airline maintenance costs. Additionally, fuel tanks support with a high strength-to-weight ratio can withstand heavy loads without excessive structural weight. It increases the fuel capacity.

2 Novel Materials and Manufacturing Technologies Take the Place of the Legacy Ones in Aviation Science

The aviation sector quickly adopts innovative materials to improve aircraft performance and capabilities. These materials significantly influence how aviation will develop, from high thrust-to-weight ratio, strength-to-weight ratio, and low buy-to-fly ratio to enhancing energy saving.

Regarding technology, the aviation sector has always been the pioneer, continuously pushing the envelope and discovering new horizons [30].

Due to continued research and development, more cutting-edge and ecologically friendly materials are expected to be incorporated into airplanes, paving the way for a more developed and effective aviation industry. This ensures lower carbon emissions and air pollution, resulting in a cleaner, greener world.

On the other hand, as mentioned at the beginning of this chapter, the aviation industry is forced to use the most advanced and sophisticated technologies to produce aircraft that operate under challenging conditions. For example, the temperature of a turbojet engine's combustion chamber is approximately 2000 °C as depicted in Fig. 13.

It can be found in the open literature that lava from Erta'Ale Lava Lake volcano in Ethiopia measured 1187 °C [33], whereas the Kilauea volcano in Hawaii

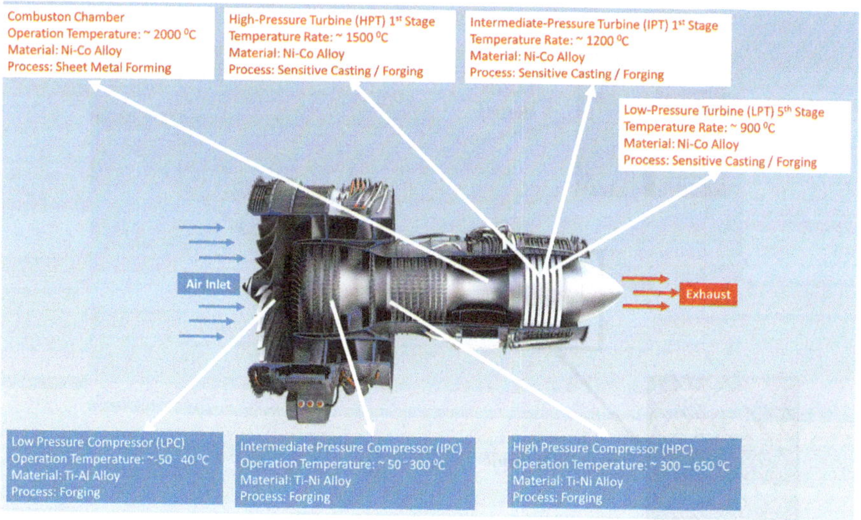

Fig. 13 The sections, materials, manufacturing process, and temperature distribution inside a turbojet engine. (Re-illustrated based on [31, 32])

measured 1230 °C [34]. Also, first stage of HPT is higher than lava from Erta'Ale Lava Lake and Kilauea volcano. It should be noted that the temperature of the combustion chamber during a regular flight operation is approximately 50% hotter than that of molten lava in volcanoes. Next, we will focus on the "high thrust-to-weight ratio, strength-to-weight ratio, and low buy-to-fly ratio" achievements in aircraft and aero engines which use innovative technologies.

2.1 Carbon Fiber Reinforced Polymers (CFRP)

CFRPs are composite materials with carbon-fibers incorporated in a polymer matrix, as shown in Fig. 14.

These materials have several important properties that make them ideal for various applications. CFRPs have the following main specifications:

- High strength-to-weight ratio: CFRP composites have remarkable strength qualities, allowing lightweight constructions to endure loads and impacts.
- Corrosion resistance: CFRP composites are highly corrosion resistant, making them appropriate for use in situations frequently exposed to moisture, chemicals, or adverse weather conditions.
- High fatigue durability: CFRP composites are extremely fatigue resistant, which means they can survive High Cyclic Loads (HCL), Low Cyclic Loads (LCL), Random Cyclic Loads (RCL), and unloading without significantly deteriorating their mechanical properties.

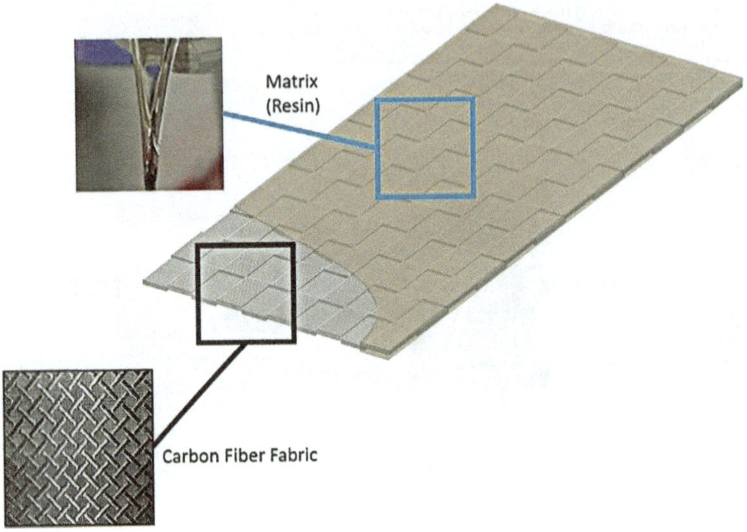

Fig. 14 Components of CFRP showing carbon-fiber fabric and matrix. (Re-illustrated based on [35])

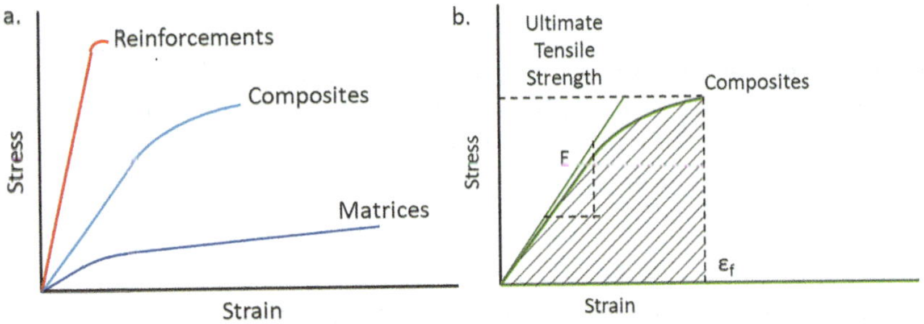

Fig. 15 (**a**) The strength qualities of reinforcement, matrices, and composites, (**b**) The physical parameters generated from a composite's stress-strain curve [36]

- High specific stiffness: CFRP composites have a high specific stiffness, which means they are stiff and unyielding in relation to their weight. This feature enables constructions to maintain shape and stability under various stress circumstances.
- Thermal conductivity: CFRP composites are thermally conductive, enabling efficient heat transport within structures.
- Excellent electrical properties: CFRP composites have outstanding electrical conductivity, making them ideal for applications requiring high electrical performance.

As shown in Fig. 15, the stress-strain graphic materials can exploit the features of the reinforcement and matrix to produce significant property gains, particularly in strength. The end-part can have enhanced mechanical features and/or new functionalities, such as improved barrier performance.

The CFRP is a synthetic substance, also called carbon fiber fabric (Kevlar) that is resistant to corrosion and five times stronger than steel [37] arises from its unique molecular structure and bonding characteristics [38].

Because of these properties, CFRPs are commonly used in the construction of various aircraft components, such as the wings, fuselage, and tail, as depicted in Fig. 16. The Boeing 787 Dreamliner and Airbus A350 XWB are made of 50% or higher composite materials, significantly reducing their weight compared to traditional aircraft.

CFRP is also considered an environmentally friendly material. It is a highly recyclable material, and the manufacturing process generates fewer emissions than other materials, such as steel and aluminum. This is consistent with the increased emphasis on sustainability in the aviation industry.

In conclusion, using CFRP in the aviation industry provides various advantages. Its lightweight nature, high strength-to-weight ratio, corrosion resistance, design versatility, and environmental friendliness make it a suitable material for aircraft manufacture. Airlines that use CFRP can save significantly on fuel usage, maintenance, and greater payload capacity.

2.2 *Additive Manufacturing (AM) Technology*

Additive manufacturing is the technique of producing an object layer by layer as opposed to subtractive (chip-away) methods. It can work with any material, from polymers to metals, and offers benefits such as process speed, customization, and

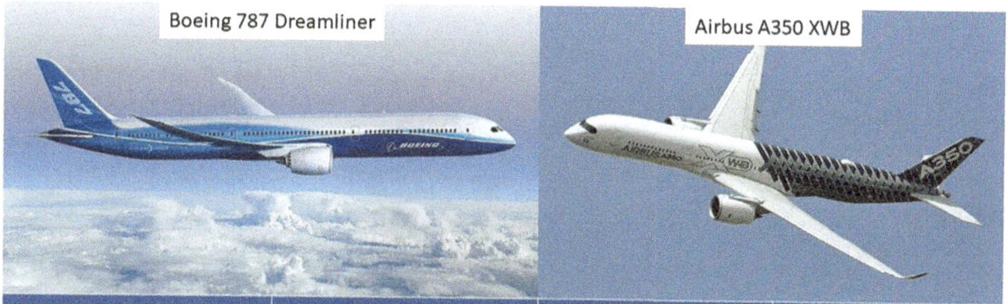

Material Type	Percentage	Material Type	Percentage
Composites (CFRP etc.)	50	Composites (CFRP etc.)	53
Aluminum – Al/li Alloy	20	Aluminum – Al/li Alloy	19
Titanium	15	Titanium	14
Steel	10	Steel	6
Miscellaneous	5	Miscellaneous	8

Fig. 16 CFRP and material distribution on Boeing 787 Dreamliner and Airbus A350 XWB aircraft usage. (Re-illustrated based on [39])

complicated geometries. This technology is still in its early stages; however, it provides incredible opportunities to decrease the final product's process time and cost.

The additive manufacturing process has aggressively expanded into the aviation manufacturing sector. The aerospace industry has relied on traditional chip removal processes for nearly a century. It can still be claimed that legacy manufacturing technologies such as milling, turning, and grinding continue to dominate the industry. These traditional machining techniques are still employed in the aerospace sector for large metal items designed especially for CNC machining processes, which use subtractive techniques, potentially resulting in up to 98% of material waste [40]. As mentioned previously, the least wastage and scrap is the best since aviation-grade materials are costly. The additive manufacturing process built a bridge to handle the lowest buy-to-fly ratio. For example, General Electric (GE)'s Leading Edge Aviation Propulsion (LEAP) fuel nozzle was additively manufactured by consolidating 20 substances into one, using a single machine, and weighing 25% less than conventionally produced fuel nozzles. GE has announced that 100.000 nozzles were additively manufactured recently [41] (Fig. 17).

The new LEAP engine saves around US$3 million annually compared to engines with additively manufactured fuel nozzles [42]. These engines power about half of the Boeing 737 and Airbus A320 fleets. Since many airliners at the global level operate commercial passenger flights with the latest version of Boeing 737 and Airbus A320, it is observed that many nozzles produced by AM are widely used at present.

Depending on the application, there are various types of additive manufacturing technologies, each with its own set of procedures, layering techniques, and equipment. As shown in Fig. 18, some of the most commonly utilized types of additive manufacturing are:

Fig. 17 GE's fuel nozzle, which is additively manufactured for LEAP engines. (Image courtesy of GE Aviation)

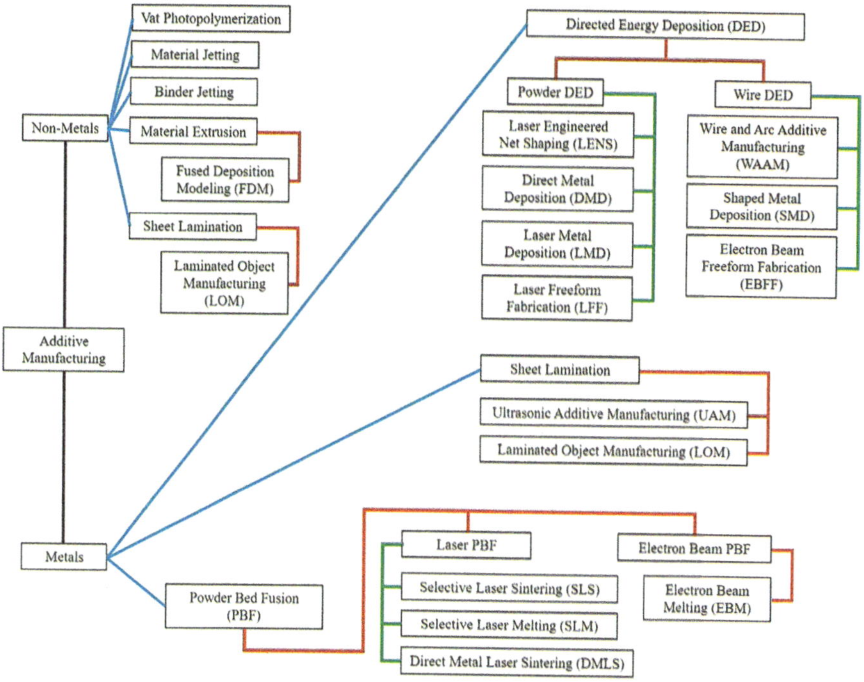

Fig. 18 The technology tree of additive manufacturing techniques. (Re-illustrated based on [24])

- Material extrusion (Fused Deposition Modeling (FDM): This process includes melting a thermoplastic material and extruding it via a nozzle to build a 3D object layer by layer.
- VAT polymerization: This method uses a vat of liquid photopolymer resin and a light source to selectively cure the resin into a solid object layer by layer.
- Powder bed fusion (PBF): This process includes spreading a thin layer of metal powder over a build platform and selectively melting the powder with a laser or electron beam to produce a 3D object layer by layer.
- Directed energy deposition (DED): This approach includes melting a metal wire or powder and depositing it on a substrate to form a 3D object.
- Binder jetting: This approach includes selectively depositing a liquid binding agent onto a bed of powder material to form a 3D object.
- Material jetting: This method includes selectively jetting material droplets onto a construction platform to generate a 3D object layer by layer.

In addition to material savings and weight reduction, it is known that safety and efficiency are the most essential features that airline operators look for in aircraft engines. The additive manufacturing technology is considered to contribute to safety and efficiency due to the use of single or multiple materials, which can positively affect the quality of the final product and the ability to control many variables such as temperature and pressure [43]. The turbofan engine used in the F-35 aircraft

(F135) features many additively made parts, as seen in Fig. 19. It's worth noting that the F135 engine's High-Pressure Turbine (HPT) and Low-Pressure Turbine (LPT) contain some essential components which are manufactured with additive manufacturing techniques. These parts are typically exposed to extremely high temperatures (~1400–1500 °C) and pressures (~20–40 Atm).

There are some important projects other than aero-engines. For example, the "SAVING" research project discovered that replacing the seat buckles utilizing AM methods can result in a 55% weight reduction compared to the original design. Weight reduction of around 72.5 kg can be achieved if all of the Airbus 380's 853 seat buckles are designed and additively manufactured. The total saves approximately 7857 tons (3.3 million gallons) of fuel over the aircraft's service life [45].

Conclusionally, thanks to the additive manufacturing process, scraps can be as low as 10%, component costs can be reduced by 50%, time-to-market can be decreased by 64%, and part weight can be lowered by 64% compared to typical manufacturing processes [20].

2.3 Graphene

Graphene is a two-dimensional material with remarkable characteristics made up of carbon atoms organized in a hexagonal lattice. It is the thinnest and most durable two-dimensional material yet known [46]. The weight of a one-square-meter graphene is 0.77 mg. It's approximately 1000 times lighter than office paper, and a

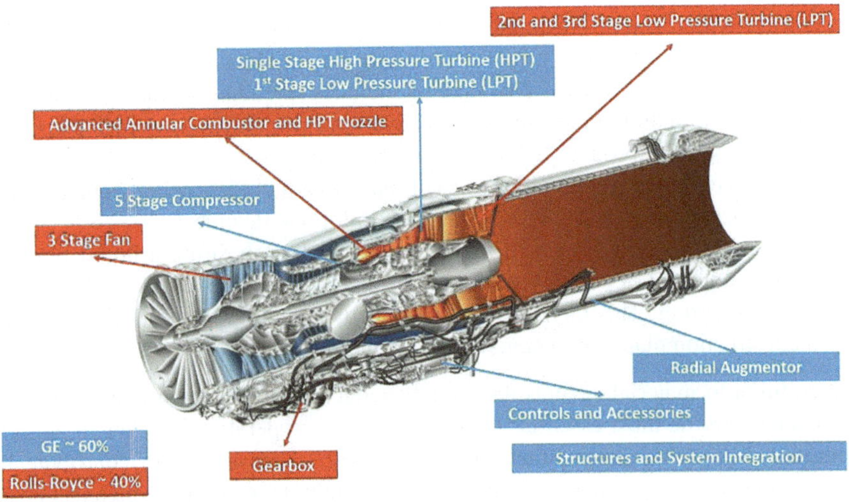

Fig. 19 The F-35 lightning's propulsion system's additively manufactured parts. (The figure was re-illustrated based on the information [44])

single sheet and one gram of graphene can cover a football field. Graphene's tensile strength exceeds 130 gigapascals [47], making it 200 times stronger than steel [48].

Because of these qualities, graphene is very advantageous to the aviation sector in several ways. For example, it can cut aircraft weight by 1%, which saves a lot of fuel and lowers CO_2 emissions. Due to its low density, it is the perfect material for building the fuselage, wings, and other parts of the aircraft, which may help drastically lower the aircraft's total weight as demanded. In this manner, a UAV named Juno was manufactured [49] using graphene-based materials in its fuselage.

The Juno aircraft, shown in Fig. 20, is a three-meter wide graphene-enhanced composite-skinned aircraft. With a wingspan of about 3.5 meters, the wing made of a graphene coating is 17% lighter than compared with carbon-fiber covering.

Juno's takeoff phase was less energy-consuming as a result of lower weight. It was observed that graphene has another significant advantage of avoiding the ice-forming on the wings. It also dealt with lightning without flinching because of the graphene-enhanced composite fuselage's high conductivity. Juno flew with graphene batteries, which do not heat up and can have more recharging capacity than its ancestors.

Coatings based on graphene can improve conductivity and provide electromagnetic shielding, protecting Juno from lightning strikes. Because of graphene's superior electrical and thermal conductivity, heat may be produced and transferred to Juno's outside surface, increasing overall efficiency. Moreover, the researchers observed that their safety and endurance increased based on graphene's great mechanical strength.

It is noteworthy to provide information about graphene-based batteries, which have five times the energy density of conventional lithium-ion batteries and are 70% lighter. Recently, NASA's X-57 Maxwell electric airplane uses graphene-based batteries [50].

Graphene's antibacterial qualities can also be used in air filtration systems, and an aircraft's Environmental Control System (ECS) can provide healthier and cleaner air to passengers and crew members. Heatable foams manufactured from

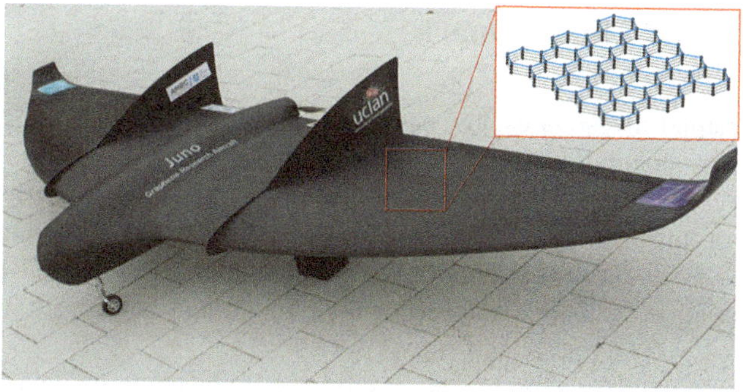

Fig. 20 Graphene-enhanced composite skinned Juno UAV flew in January 2017. (The figure was re-illustrated based on the information [49])

low-density aerographene are used in ECS, provide smart self-cleaning functions, and reduce maintenance costs.

Depending on the altitude and OAT, ice accumulation on the wing is an ordinary situation during flights. As these accumulations reach a certain point where the lift generation is disturbed and becomes a fatal flight safety issue. Some electrical heaters and rubber balloons on the leading edge of the wings are the current solutions provided by the existing technological level. Researchers were focusing on the more efficient solutions for de-icing systems. A recent study by Niu et al. [51] claims that laminated Laser-Induced Graphene (LIG) has a de-icing capability that can be used in aviation. They extended their expectations that LIG will pave a new way of de-icing. In another study by Chen et al. [52], Hydrophobic Conductive Graphene (HCG), with its exceptional electrical conductivity, is a very strong candidate for use in de-icing systems. These two studies and many more claim that graphene-based de-icing materials will be implemented into the aviation industry in a very short period.

Enhanced Lightning Strike Protection (ELSP) is a vital system that ensures flight safety while unloading the high-voltage electric current from the aircraft body. Today, composites incorporating copper mesh are mainly used in ELSP, a complex system that adds unwanted weight to aircraft. As mentioned before, decreased weight means increased payload and extended range. Also, since the electricity tends to flow along the most conductive pathway, the anisotropic conductive structure can induce the lightning current path over the aircraft. With this regard, graphene has a vital usage area in ELSP with its isotropic structure, extreme conductivity, and extreme lightness. The study accomplished by Raimondo et al. [53] underlines the flame-retardant ability of graphene, which represents a peremptory aspect in the current aerospace industry. The stated ability is essential for ELSP. Another study by Guo et al. [54] empowers the idea of graphene usage in the ELSP system.

Conclusionally, graphene is a perfect material that combines extreme conductivity, strength, lightness, and flexibility. The mentioned exceptional features will have a direct impact on flight safety and also save energy.

3 Results and Final Remark

This book chapter investigates how new materials and procedures may improve an aircraft's environmental performance and energy efficiency. Additive manufacturing technology, CFRP, and graphene are a few of the innovative materials and techniques that promise to lower aircraft weight, waste, and fuel consumption while enhancing their durability, strength, and functionality. These developments can lessen the aviation industry's impact on the climate and help meet sustainability targets.

The aviation industry's top priority is ensuring flight safety while also recognizing the need to reduce costs due to fierce competition among airliner companies. By optimizing flight operations and reducing DOC, the industry can continue to provide safe and reliable air travel to passengers worldwide. In order to reduce DOC, essential parameters such as thrust-to-weight ratio, buy-to-fly ratio, and strength-to-weight ratio must always be considered.

During the aircraft design phase, incorporating the concept of thrust-to-weight ratio leads to energy savings. By accounting for this ratio, aircraft can operate at a higher level with less energy input, which increases fuel efficiency. This, in turn, not only reduces the DOC but also decreases the adverse environmental impact.

Another essential factor that must be considered regarding energy saving in aviation is the buy-to-fly ratio. This ratio assesses how much a component weighs after it has been produced based on its raw material weight. The buy-to-fly ratio can be significantly decreased using lightweight materials like CFRP and graphene, ultimately leading to less material waste and lower manufacturing energy consumption.

Lastly, the strength-to-weight ratio, which affects an aircraft's performance and economy, is vital to energy saving. The strength-to-weight ratio can be optimized by choosing strong, lightweight materials, which enables more efficient energy use and an overall reduction in fuel consumption.

Considering these three ratios, this book chapter investigated some cutting-edge technologies and materials, including additive manufacturing, carbon-fiber-reinforced material, and graphene. In summary, the combination of these new materials and processes offers promising opportunities for energy savings in the aviation industry.

Additive manufacturing, broadly called 3D printing, revolutionized the manufacturing process. CFRP is another crucial material utilized in the aviation industry to lower weight and increase fuel efficiency. To improve energy efficiency further, the usage of graphene, a 2D material with remarkable strength and conductivity in aircraft, is being investigated. The aviation industry can now attain higher strength-to-weight ratio, thrust-to-weight ratio, and lower buy-to-fly ratio thanks to these cutting-edge technologies, which enable aircraft operating companies to make more cost-effective aerial operations. Along with cutting expenses and increasing productivity, these innovations are also assisting the sector in lessening its environmental effects. Further study and development will be necessary to address the technological, financial, and regulatory obstacles these materials and procedures must also overcome.

Acknowledgments The author thanks TUSAS for the contributions and support of the research in the aviation field to the article. He also appreciates Mr. Onur Emre Demirel's assistance in preparing Figs. 14 and 20.

Data Availability Statements The datasets used or analyzed during the current study are available from the corresponding author upon reasonable request.

Competing Interests The author states that no known competing financial interests or personal relationships could have appeared to influence this proceeding paper.

References

1. Saptarshi, S. M., & Zhou, C. (2019). Basics of 3D printing: Engineering aspects. In M. Dipaola (Ed.), *3D printing in Orthopaedic surgery*. ISBN 978-0-323-58118-9.
2. Noh, H. M., Benito, A., & Alonso, G. (2016). Study of the current incentive rules and mechanisms to promote biofuel use in the E.U. And their possible application to the çivil aviation sector. *Transportation Research Part D: Transport and Environment, 46*, 298–316. https://doi.org/10.1016/j.trd.2016.04.007
3. Ma, F., Cao, W., Luo, Y., & Qiu, Y. (2016). The review of manufacturing technology for aircraft structural part. *Procedia CIRP, 56*, 594–598. https://doi.org/10.1016/j.procir.2016.10.117
4. Centreforaviation Website. *Global air travel growth faces limitation from slow aircraft deliveries*. Retrieved January, 12, 2024. https://centreforaviation.com/analysis/reports/global-air-travel-growth-faces-limitation-from-slow-aircraft-deliveries-669289
5. Statista Website. (2024). *Worldwide number of Airbus aircraft deliveries from 2002 to 2021. Retrieved January, 12.* https://www.statista.com/statistics/264493/airbus-worldwide-aircraft-deliveries/
6. Chestnotes Website. *Boeing commercial airplane deliveries*. Retrieved January, 12, 2024. https://cheshnotes.com/boeing-commercial-airplane-deliveries/
7. ICAO. (2016). Capacity and Efficiency Report, 2016–2030 Global Air Navigation Plan, 6, ISBN: 978-92-9258-000-1, from https://www.icao.int/airnavigation/documents/ganp-2016-interactive.pdf
8. Wise, M., Muratori, M., & Kyle, P. (2017). Biojet fuels and emissions mitigation in aviation: An integrated assessment modeling analysis. *Transportation Research Part D: Transport and Environment, 52*, 244–253. https://doi.org/10.1016/j.trd.2017.03.006
9. Jones, G., Ryan, E. T., Ritschel, J. D. (2014). Investigation into the Ratio of Operating and Support Costs to Life-Cycle Costs for DoD Weapon Systems", Defense ARJ, Vol. 21, no. 1, pp. 442–464, 2014.
10. Sher, F., Raore, D., & Klemeš, J. J. (2021). Unprecedented impacts of aviation emissions on global environmental and climate change scenario. *Current Pollution Reports, 7*, 549–564. https://doi.org/10.1007/s40726-021-00206-3
11. Gelhausen, M. C., Berster, P., & Wilken, D. (2018). A new direct demand model of long-term forecasting air passengers and air transport movements at German airports. *Journal of Air Transport Management, 71*, 140–152. https://doi.org/10.1016/j.jairtraman.2018.04.001
12. Fioriti, M., Vercella, V., & Viola, N. (2018). Cost-estimating model for aircraft maintenance. *Journal of Aircraft, 55*, 1–12. https://doi.org/10.2514/1.C034664
13. Kaufmann, M. (2008). Cost/weight optimization of aircraft structures, licence thesis, KTH School of Engineering Sciences, ISBN 9789171788887, Stockholm, Sweden.
14. Justin, H. (2006). Boeing 787 from the ground up. *Boeing Aero Magazine*, 21.
15. Gorbatikh, L., Wardle, L., & B., Lomov, S. V. (2016). Hierarchical lightweight composite materials for structural applications. *Cambridge MRS Bulletin, 41*(9), 672–677. https://doi.org/10.1557/mrs.2016.170
16. Steinegger, R. (2017). Fuel economy as function of weight and distance. *Zurcher Fachhochschule*. Retrieved January, 14, 2024, Retrieved January, 12, 2024. Available online: https://digitalcollection.zhaw.ch/bitstream/11475/1896/6/Steinegger_Fuel_Economy_as_a_Function_of_Weight_and_Distance_v1-1.pdf
17. Ranasinghe, K., Guan, K., Gardi, A., & Sabatini, R. (2019). Review of advanced low-emission technologies for sustainable aviation. *Energy, 188*, 115945. https://doi.org/10.1016/j.energy.2019.115945
18. Biasco, R. (2021). Emissions analysis routine for subsonic aircrafts using biofuel, Politecnico di Torino, M. Sc. thesis.
19. Tsai, W. H., Chang, Y. C., Lin, S. J., Chen, H. C., & Chu, P. Y. (2014). A green approach to the weight reduction of aircraft cabins. *Journal of Air Transport Management, 40*, 65.

20. Deloitte University. (2014). *3D opportunity in aerospace and defense: Additive manufacturing takes flight*, 4, (2014).
21. Ariffin, L. M., Rostam, A. H., & Shiban, W. M. E. (2019). Study of aircraft thrust-to-weight ratio. *Journal of Aviation and Aerospace Technology, 1*(2), 1–9.
22. Davis, L. A. (2016). First stage recovery. *Engineering, 2-2*, 152–153. https://doi.org/10.1016/J.ENG.2016.02.007
23. Barz, A., Buer, T., & Haasis, D. (2016). A study on the effects of additive manufacturing on the structure of supply networks. *International Federation of Automatic Control, 4*. https://doi.org/10.1016/j.ifacol.2016.03.013
24. Yusuf, S. M., Cutler, S., & Gao, N. (2019). Review: The impact of metal additive manufacturing on the aerospace industry. *Metals., 9*, 1286. https://doi.org/10.3390/met9121286
25. Dehoff, R., Tallman, C., Duty, C., Peter, W., Yamamoto, Y., Chen, W., & Blue, C. (2013). Case study: Additive manufacturing of aerospace brackets. *Advanced Materials and Processes., 171*. https://doi.org/10.31399/asm.amp.2013-03.p019
26. Ford, S., & Despeisse, M. (2016). Additive manufacturing and sustainability: An exploratory study of the advantages and challenges. *Journal of Cleaner Production, 137*, 1573–1587. https://doi.org/10.1016/j.jclepro.2016.04.150
27. Gisario, A., Kazarian, M., Martina, F., & Mehrpouya, M. (2019). Metal additive manufacturing in the commercial aviation industry: A review. *Journal of Manufacturing Systems, 53*, 124–149., ISSN 0278-6125. https://doi.org/10.1016/j.jmsy.2019.08.005
28. Sinha, K., Klimmek, T., & Schulze, M. (2021). Loads analysis and structural optimization of a high aspect ratio, composite wing aircraft. *CEAS Aeronautical Journal, 12*, 233–243. https://doi.org/10.1007/s13272-021-00494-x
29. Li, M., Yuan, J., & Guan, D. (2011). Application of piezoelectric fiber composite actuator to aircraft wing for aerodynamic performance improvement. *Science China Technological Sciences., 54*, 395–402. https://doi.org/10.1007/s11431-010-4212-0
30. Bednarska, K., Sobotka, P., Woliński, T. R., Zakręcka, O., Pomianek, W., Nocoń, A., & Lesiak, P. (2020). Hybrid fiber optic sensor systems in structural health monitoring in aircraft structures. *Materials, 13*(10), 2249. https://doi.org/10.3390/ma13102249
31. Spittle, P. (2003). Gas turbine technology. *Physics Education, 38*(6), 504–511.
32. Mitsuhiro, T., & Masashi, K. (2014). Making lighter aircraft engines with titanium aluminide blades. *IHI Engineering Review, 47*(1), 10–13.
33. Burgi, Y. P., Caillet, M., & Haefeli, S. (2002). Field temperature measurements at Erta'Ale lava Lake, Ethiopia. *Bulletin of Volcanology, 64*, 472–485. https://doi.org/10.1007/s00445-002-0224-3
34. Radebaugh, G. T., Saito, J. T., Lorenz, R. D., Dangerfield, A., & Tingey, D. G. (2015). Temperatures, thermal structure, and behavior of eruptions at Kilauea and Erta Ale volcanoes using a consumer digital camcorder. *GeoResJ, 47–57*. https://doi.org/10.1016/j.grj.2015.01.001
35. Khan, I. N., & Sudipta, H. (2020). Chapter 15 – Self-healing fiber-reinforced polymer composites for their potential structural applications. In T. Sabu & S. Anu (Eds.), *Self-healing polymer-based systems* (p. 455). Elsevier. https://doi.org/10.1016/B978-0-12-818450-9.00015-5
36. Miao, C., & Hamad, W. Y. (2013). Cellulose reinforced polymer composites and nanocomposites: A critical review. *Cellulose, 20*, 2221–2262. https://doi.org/10.1007/s10570-013-0007-3
37. Uyar, I. (2014). *Experimental investigation of dynamic delamination in curved composite laminates*. M. Sc. Thesis, Middle East Technical University.
38. Grujicic, M., Bell, W., Glomski, P., Pandurangan, B., Yen, C. F., & Cheeseman, B. (2011). Filament-level modeling of aramid-based high-performance structural materials. *Journal of Materials Engineering and Performance., 20*, 1401–1413. https://doi.org/10.1007/s11665-010-9786-y
39. Bazli, M., & Bazli, L. (2021). A review of the mechanical properties of synthetic and natural fiber-reinforced polymer composites and their application in the transportation industry. *Journal of Composites and Compounds, 3*. https://doi.org/10.52547/jcc.3.4.6

40. Allen, J. (2006). *An investigation into the comparative costs of additive manufacture vs. machine from solid for aero engine parts*. Manufacturing Technology Rolls-Royce plc, Unclassified Report, United Kingdom.
41. Giannotta, N., Sala, G., Bianchini, C., & Torreggiani, A. (2023). A review of additive manufacturing of soft magnetic materials in electrical machines. *Machines, 11*, 702. https://doi.org/10.3390/machines11070702
42. Knofius, N., Van der Heijden, M. C., & Zijm, W. H. M. (2018). Consolidating spare parts for asset maintenance with additive manufacturing. *International Journal of Production Economics*, (), S0925527318304468–. https://doi.org/10.1016/j.ijpe.2018.11.007.
43. Rizzo, A., Goel, S., Luisa, G. M., Iglesias, R., Jaworska, L., Lapkovskis, V., Novak, P., Postolnyi, B. O., & Valerini, D. (2020). The critical raw materials in cutting tools for machining applications: A review. *Materials, 13*, 1377. https://doi.org/10.3390/ma13061377
44. Froes, F., Boyer, R., & Dutta, B. (2019). Introduction to aerospace materials requirements and the role of additive manufacturing. In *Additive manufacturing for the aerospace industry*. 1–6, p. 482, Elsevier, Amsterdam, ISBN:78-0-12-814062-8.
45. Singamneni, S., Yifan, L. V., Hewitt, A., Chalk, R., & Thomas, W. (2019). Additive manufacturing for the aircraft industry: A review. *Journal of Aeronautics & Aerospace Engineering, 8*(214), 2. https://doi.org/10.4172/2329-6542.1000214
46. Kumar, Y. B., Ravindra, K. R., Ashutosh, T. G., & Narahari, S. (2022). Chapter 7 - Reversible and irreversible functionalization of graphene. In T. Dinadayalane & F. Hagelberg (Eds.), *Theoretical and computational chemistry* (Vol. 21, pp. 157–189., ISSN 1380-7323, ISBN 9780128195147). Elsevier. https://doi.org/10.1016/B978-0-12-819514-7.00005-1
47. Nassef, B., Nassef, G., & Daha, M. A. D. E. (2020). Graphene and its industrial applications–A review. *International Journal of Materials Engineering, 10*, 1–12. https://doi.org/10.5923/j.ijme.20201001.01
48. Sarkar, I. J. R., Badiyani, A. D., Sonagara, L. B., Keralia, S. L., Chudasama, N., & Kumar, S. (2023). Chapter 1- Graphene-based nanomaterials. In S. K. Patel, S. A. Taya, S. Das, & K. V. Babu (Eds.), *Recent advances in graphene nanophotonics*. Springer. https://doi.org/10.1007/978-3-031-28942-2
49. Khalid, Y. K., Kamal, A., Otabil, A., Mamoun, O., & Liao, K. (2023). Graphene/epoxy nanocomposites for improved fracture toughness: A focused review on toughening mechanism. *Chemical Engineering Journal Advances, 16*, 100537. https://doi.org/10.1016/j.ceja.2023.100537
50. Bright, A. A. G., & Good, C. (2022). Electric aviation: A review of concepts and enabling technologies, transportation. *Engineering, 9*, 100134. https://doi.org/10.1016/j.treng.2022.100134
51. Niu, Y., Sun, Y., Tao, X., Liao, C., & Yao, J. (2024). Superhydrophobic-electrothermal laser-induced graphene for effectively anti-icing and de-icing. *Composites Communications, 101830*. https://doi.org/10.1016/j.coco.2024.101830
52. Chen, J., Marklund, P., Björling, M., & Shi, Y. (2023). In-situ polymerized siloxane urea enhanced graphene-based super-fast, durable, all-weather elec-photo-thermal anti–/de-icing coating. *Journal of Science: Advanced Materials and Devices, 8*(3). https://doi.org/10.1016/j.jsamd.2023.100604
53. Raimondo, M., Guadagno, L., Speranza, V., Bonnaud, L., Dubois, P., & Lafdi, K. (2018). Multifunctional graphene/POSS epoxy resin tailored for aircraft lightning strike protection. *Composites Part B: Engineering, 140*, 44–56. https://doi.org/10.1016/j.compositesb.2017.12.015
54. Guo, Y., Xu, Y., Wang, Q., Dong, Q., Yi, X., & Jia, Y. (2019). Enhanced lightning strike protection of carbon fiber composites using expanded foils with anisotropic electrical conductivity. *Composites Part A: Applied Science and Manufacturing, 117*, 211–218. https://doi.org/10.1016/j.compositesa.2018.11.022

Biocomposites as Aircraft Materials

Norkhairunnisa Mazlan, Ain Umaira Md Shah, Tay Chai Hua, Izzat Najmi, and Natasha Ramli

1 Introduction

The Wright Brothers created the first powered aircraft in 1903 using wood, wire, and fabrics made of silk or cotton. Wood from red spruce trees is chosen for its lightweight features, while ash wood is chosen due to its good bending properties. The procurement of solid wood for aircraft necessitates strict adherence to rigorous

N. Mazlan (✉)
Department of Aerospace Engineering, Faculty of Engineering, Universiti Putra Malaysia, Serdang, Selangor, Malaysia

Aerospace Malaysia Research Center (AMRC), Faculty of Engineering, Universiti Putra Malaysia, Serdang, Selangor, Malaysia

Institute of Nanoscience and Nanotechnology (ION2), Universiti Putra Malaysia, Serdang, Selangor, Malaysia
e-mail: norkhairunnisa@upm.edu.my

A. U. Md Shah · I. Najmi
Department of Aerospace Engineering, Faculty of Engineering, Universiti Putra Malaysia, Serdang, Selangor, Malaysia

Aerospace Malaysia Research Center (AMRC), Faculty of Engineering, Universiti Putra Malaysia, Serdang, Selangor, Malaysia
e-mail: ainumaira@upm.edu.my

T. C. Hua · N. Ramli
Department of Aerospace Engineering, Faculty of Engineering, Universiti Putra Malaysia, Serdang, Selangor, Malaysia

Aerospace Malaysia Research Center (AMRC), Faculty of Engineering, Universiti Putra Malaysia, Serdang, Selangor, Malaysia

Institute of Tropical Forestry and Forest Product (INTROP), Universiti Putra Malaysia, Serdang, Selangor, Malaysia

T. H. Karakoc et al. (eds.), *Sustainable Materials and Manufacturing Techniques in Aviation*, Sustainable Aviation, https://doi.org/10.1007/978-3-031-62987-7_2

standards and stringent quality requirements based on aircraft authorities. Advantages of wood includes ease of manufacture and lightweight with a high strength-to-weight ratio that allows for substantial width to give lateral stability against buckling. Nevertheless, certain limitations exist associated with using wood in the context of aircraft. Wood degrades rapidly which is attributed to its hygroscopic properties, causing it to quickly absorb moisture, particularly when subjected to prolonged exposure in a wet environment. The biological characteristics of wood, such as variations in grain structure, inhomogeneity features of the fibres, and ease of splintering and buckling, require a lot of maintenance when using wood as aircraft structure [1]. Hence, although wood has historically been employed in aircraft construction since the inception of aviation, there has been a surge of interest in utilizing biocomposite materials for these limitations recently.

The biocomposite material is well-known for its inherent characteristics of biodegradability and sustainability. The material is of great interest since it can be recycled or disposed of at the end of its lifespan without negatively affecting the environment. By 2030, the market for biocomposite is expected to grow by a CAGR of roughly 14.8%, from USD 24,726.7 million in 2022 to USD 74,593.5 million [2]. The use of composite materials in aircraft construction has been shown to have the potential to decrease fuel consumption due to its advantageous lightweight characteristics. Additionally, adopting biocomposites can contribute to reducing carbon dioxide (CO_2) emissions [3, 4]. Furthermore, natural fibres are inexpensive, lightweight, possess complex chemical and biological characteristics, and are sustainable. Gomez-Campos et al. [5] examined the weight characteristics of biocomposites compared to standard synthetic fibre-reinforced polymer composites. Their findings indicate that biocomposite exhibits a slightly greater weight. Variations in weight were seen because of the inherent high porosity of the natural fibre, which leads to a greater absorption of resin. Therefore, modifying the biocomposite design could mitigate the concern of excessive mass. By maintaining an equivalent weight to typical polymer composites, it is possible to decrease the emission of CO_2 into the environment. Despite the relatively modest contribution of fossil fuel-related greenhouse gas (GHG) emissions in the aviation industry, accounting for approximately 2.4%, there is a discernible escalation in public concern regarding the prevailing climate change scenario [6].

Biocomposite is a composite material formed by combining natural fibres with either bioresin or synthetic polymers. Biocomposite-based products have the potential to offer viable solutions for challenges related to petroleum scarcity and waste management. The material provides numerous benefits, including its capacity to biodegrade, ease of processing, affordability, lightweight nature, and sustainability. Extensive research and practical implementation have been conducted on the use of biocomposite materials in aircraft applications, particularly in the fabrication of anisotropic and lightweight structural components such as aircraft panels. These materials have found significant application in the interior and secondary structures of aircraft, where the stress levels experienced by these components are comparatively less critical [7]. Nevertheless, it is essential to acknowledge that natural fibres possess certain limitations. These drawbacks, such as high moisture absorption, which increases the risk of damage and poor wettability, which may lead to delamination due to mismatched surface properties with polymers. By considering the drawbacks of natural fibre, it is essential to explore the technology to improve the natural fibre's properties, such as chemical treatment or surface modifications. This is because the adherence of biocomposite materials to safety and strict regulations specified by aviation authorities is of utmost importance to uphold aircraft safety and mitigate accidents or mishaps.

The development of environmentally friendly composites for aircraft is essential for a sustainable society, particularly considering the current global environmental changes. The development of biocomposite materials for aircraft aligns with several Sustainable Development Goals (SDGs) outlined by the United Nations. Notably, these goals include SDG 7 (Affordable and Clean Energy), SDG 8 (Decent Work and Economic Growth), SDG 9 (Industry, Innovation, and Infrastructure), SDG 12 (Responsible Consumption and Production), and SDG 13 (Climate Action). The Sustainable Development Goals (SDGs) prioritize the promotion of resource efficiency, the reduction of waste generation and its potential utilization, the limited use of hazardous substances, and the establishment of a harmonious relationship between humans and the natural environment to attain a society with low carbon emissions [8, 9].

Another factor driving the push for enhanced development and use of biocomposites in the aviation industry is the challenges associated with the disposal of panels and components from decommissioned aircraft. For instance, 50% of the Boeing 787's body structure is made up of carbon fibre-reinforced polymer (CFRP) [10], and it is expected that by 2050, about 500,000 tonnes of CFRP waste will accumulate at the end-of-life phase of the aircraft [11]. CFRP is an engineered composite material with exceptional attributes, making it an optimal choice for aircraft structures. Its lightweight nature, remarkable strength, stiffness, and fatigue resistance render CFRP very suitable for aircraft bodies [12]. According to the International Air Transport Association (IATA) [13], about 700 aircraft reach the end of their operational lives every year, and the trend will keep increasing with more waste generated. Thus, with a high volume of CFRP in the aircraft body, composite recycling has become an integral part of aircraft manufacturers' ambitions for sustainable aviation. Here comes the challenge of reusing and recycling the CFRP materials, or they will end up in landfills. Thus, an alternative to reduce the environmental concern and cost of recycling or reusing CFRP composite waste is by utilising biocomposite materials. An investigation by Ramachandran et al. [14] indicated that carbon fibre composites had approximately five times more environmental impact than flax-based biocomposite materials in emission impact analysis.

Given the constraints imposed by the finite supply of non-renewable petrochemical-based resources to produce high-strength synthetic fibres and advanced polymeric materials, biocomposites are increasingly being recognized as a promising substitute material within the aerospace sector. The focus of this chapter is categorizing biocomposites, examining aircraft parts and components that can be manufactured using biocomposites, and discussed significant analyses that have been conducted to evaluate the suitability of biocomposite for aircraft structure applications. Furthermore, recent studies pertaining to the sustainability aspects of biocomposites in the context of aircraft were also explored. Future research direction to ensure the sustainability of biocomposites are discussed as well. Lastly, the chapter recommends the steps of advanced end of life cycle treatment system for biocomposites.

2 Single and Hybrid Biocomposites

Multiple definitions of biocomposites have been documented. Biocomposites are composite materials that combine natural reinforcement and polymer matrix [15]. As previous studies [16, 17] have highlighted, it is significant to note that the reinforcement or the polymer phase in biocomposites originates from a natural source. However typically, the natural component in biocomposites is the reinforcement, known as Natural Fibre Reinforced Polymers [18].

In recent years, there has been a significant amount of research focused on investigating the hybridization of natural and synthetic fibre reinforcement [19] as well as the hybridization of natural and synthetic polymer, commonly referred to as polymer blend [20]. This is attributed to their enhanced performance capabilities, rendering them more suitable for a particular application. Therefore, a diverse range of materials can be classified as biocomposites. Figure 1 presents the classification of biocomposites according to previously established definitions. Hence, within the framework of this article, the inclusion of any reinforcement or polymer derived from natural sources, or a combination of both, will be referred to as biocomposites and will be the subject of discussion in this chapter.

2.1 Current Composites Used in Aircraft

Composites have made their way into the structures of aircraft, particularly the interior parts such as ceilings, sidewalls, furniture, floor, and seating [21]. High-strength fibres such as glass, carbon, and aramid are used in these composites [22]. As of 2023, the Federal Administration Aviation has officially acknowledged using composites in aircraft, as indicated in Table 1. As mentioned earlier, the composites adhere to the flammability standards outlined in FAR Part 25. Generally, these materials used are the same regardless of the type of aircraft involved, either normal or transport category aircraft and/or rotorcraft [23]. These composites can be a good baseline for producing biocomposites that aim to be used in aircraft. Table 1 can assist researchers in effectively selecting appropriate potential applications for their created biocomposites.

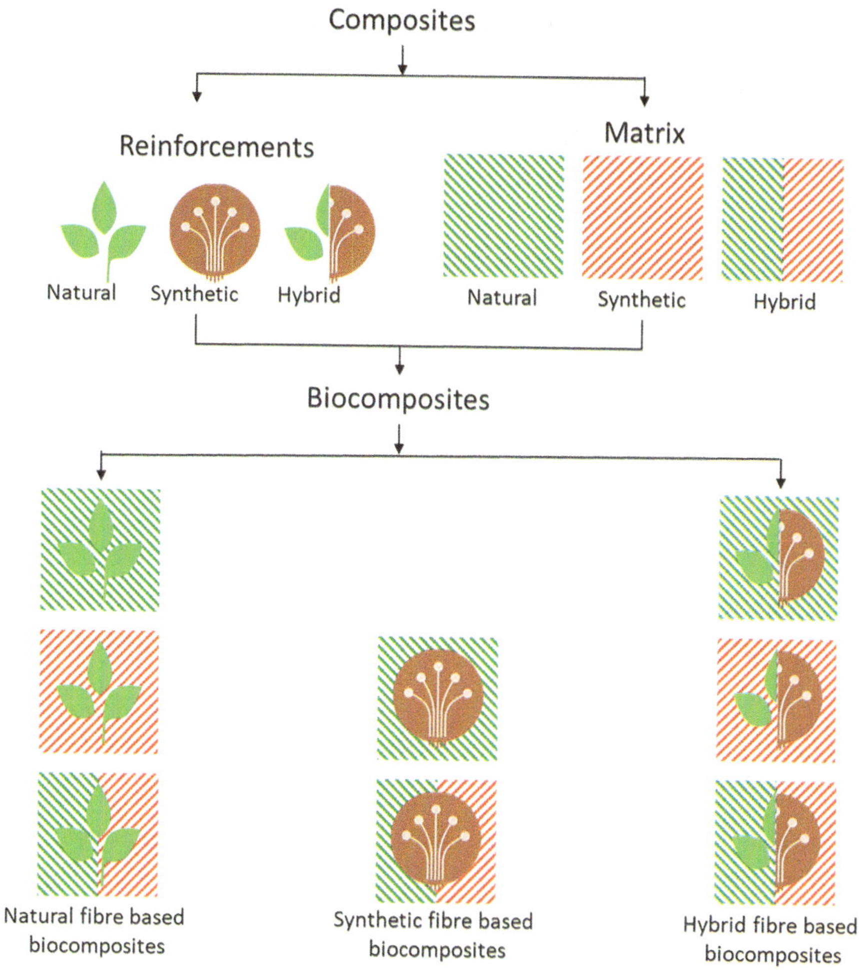

Fig. 1 Classification of biocomposites based on fibre reinforcements

2.2 Biocomposites Studied for Aircraft Parts

This section exclusively presents biocomposites that have been documented as having possible uses in the field of aircraft. These biocomposites are shown in Table 2.

2.2.1 Advanced Biocomposite

According to Wu et al. [34], advanced fibres represent a novel iteration of fibres that possess several notable attributes, including the ability to be produced on a large scale, the capacity to retain certain features and a diverse range of potential uses.

Table 1 Parts in aircraft interiors that are made of composites [23]

Aircraft parts	Components	Composites
Aircraft seat	Structure	Carbon composites
Insulating materials	Batting	Fibrous glass reinforced phenolic
	Engine pylons, nacelles, power units and engine air ducting	Fibrous glass reinforced silicone binder
	Insulation cover	Nylon yarn reinforced polyester/polyvinyl fluoride
Interior panel structure	Ceilings, galleys, lavatories, sidewalls, baggage racks, floor, partitions, and closets	Fibre reinforced resin
Floor covering	Areas where fluid likely spills such as galleys and lavatories	Wool/nylon yarns reinforced polyester/polypropylene
		Fabric reinforced vinyl
Non-metallic air ducting	Duct boots	Woven fibreglass with polyester/epoxy/phenolic resin/aromatic polyamide
		Fibreglass impregnated with silicone rubber
Linings (nonpanel)	Passenger and food cart	Fibreglass reinforced resins
Firewalls	Even more burn through resistance in specific areas	Silica glass reinforced resin

Advanced biocomposites are composite materials that integrate natural fibres or matrices obtained from biological sources and employ advanced engineering techniques and additives to improve their performance and properties. The materials have been specifically formulated to fulfil designated performance criteria and adhere to sustainability objectives. To address the limitations of natural fibres, several treatments are employed to enhance their qualities and promote compatibility with the polymer matrix, and to enhance its flame retardancy [22]. The utilization of flame retardant-treated woven flax fabric has been explored in the manufacture of biocomposites intended for commercial cabin interior panels. The treated flax is then reinforced with phenolic resin. The present phase of the project focuses on the initial stages of the programme, which aims to facilitate the creation of biocomposites comprising flax that have been processed and combined with a biopolymer. The intended application for these biocomposites is within the aerospace industry, as highlighted by John et al. [24].

Other natural fibre weaknesses include non-uniformity along the length, resulting in several weak links [35] and high sensitivity to thermal damage [36]. This limitation hinders replacing synthetic fibres with natural fibres for use in aircraft components [37]. To counter this, natural fibres can be hybridized with synthetic fibres. At present, this approach is in the process of being prototyped and validated. However, this is seen as a faster means to market than fully substituting synthetic fibres with natural fibres for aerospace parts [38].

Table 2 The biocomposites reported for potential aircraft parts

Fibre			Polymer			
Natural						
Type	Treatment	Synthetic	Natural	Synthetic	Aircraft part	Ref
Kenaf	NaOH	Glass	–	Bisphenol A diglycidyl ether (DGEBA)	Radome	[22]
Flax	Flame retardant	–	–	Phenolic resin	Panels for cabin	[24]
Hemp	–	–	–	DGEBA	Wing	[25]
Flax						
Sisal						
Ramie						
Hemp	–	–	Poly-lactic-acid (PLA)	–		
Flax						
Sisal						
Ramie						
Jute	–	–	Glucofuranoside-based trifunctional EP monomer (GTFE)	–	Sandwich composite for aircraft interior flooring	[26]
Carbon						
Jute	–	–	–	DGEBA		
				Aliphatic Trifunctional Glycerol (GER)		
				Tetrafunctional Pentaerythritol based (PER)		
Flax				DGEBA	Tailplane and cabin door of an ultralight helicopter	[27]
Flax				DGEBA	Aircraft interior like partition panels, tray tables, and baggage compartments	[7]
Kenaf				Acrylonitrile butadiene styrene (ABS)	Aircraft secondary structure	[28]
Ramie				DGEBA	Aircraft secondary structure such as overhead locker	[29]

(continued)

Table 2 (continued)

Fibre			Polymer			
Natural						
Type	Treatment	Synthetic	Natural	Synthetic	Aircraft part	Ref
Hybridized banana and sisal				DGEBA	Not mentioned	[30]
Rice Husk	Silane-treated			DGEBA	Air-duct	[31]
Cotton	–	–	Castor oil–based polyurethane	–	Aircraft interior structures	[32]
Glass						
Cotton	–	–	–	DGEBA		
Glass	–	–	Castor oil–based polyurethane	–	Aircraft interior structures	[33]
Cotton						
Cotton	–	–	–	DGEBA		

High-performance polymers refer to a class of polymers that exhibit exceptional characteristics, as they can maintain their ideal mechanical, thermal, and chemical capabilities even in the presence of challenging conditions, such as extreme temperatures, high pressures, and corrosive chemicals [39]. The demand for high-performance polymers in aircraft applications necessitating strict material requirements is substantial [40]. The ongoing shift from synthetic to natural polymers has prompted the exploration of high-performance biopolymers, primarily focusing on the epoxidation of mineral oil-based materials [41]. In recent times, there have been multiple efforts to identify a suitable alternative for the common thermoset composite matrix material, namely the diglycidyl-ether of -bisphenol -A (DGEBA) epoxy monomer [42]. In aerospace applications, four distinct categories of biopolymers have been identified as having the ability to be utilized as alternatives to synthetic polymers. These biopolymers include poly-lactic acid (PLA), glucofuranoside-based trifunctional EP monomer (GTFE), and castor oil-based Polyurethane (PU).

Of the four biopolymers described, it is noteworthy to notice the features of the GTFE polymer. When treated with amine-type diethylene-toluene-diamine (DETDA), the jute fibre-reinforced biocomposite based on GTFE exhibits a significantly higher glass transition temperature (T_g) compared to biocomposites reinforced with aliphatic trifunctional glycerol (GER) and tetrafunctional pentaerythritol (PER). Specifically, the T_g of the GTFE-based biocomposite is 122% and 157% higher than that of GER and PER, respectively. However, it is worth noting that the T_g of the GTFE-based biocomposite is 6% lower than that of a mineral oil-based aromatic bifunctional DGEBA-based biocomposite. A comparable pattern is observed when carbon fibre is utilized as reinforcement for polymers. The mechanical testing of Tensile Strength, Young's Modulus, Bending Strength, and Bending Modulus revealed that the carbon fibre reinforced GTFE polymer exhibited superior values compared to the DGEBA polymers. These results indicate that the newly developed biopolymer has the potential to serve as a viable alternative to synthetic materials in aircraft applications [26].

3 Fibre Orientation

The orientation of the reinforcing material additionally influences the properties of fabricated composites. The fibre orientation determines the isotropic and anisotropic behaviour of composites. The selection of composites is determined by the specific loading conditions, considering the fibres' orientation. According to Amor et al. [42], for one-directional loading, such as tension and compression applications, the arrangement or orientation of fibre must be in the direction of load, demonstrating the highest load-bearing capabilities of the material. In addition, the material can exhibit both isotropic and anisotropic mechanical behaviour in applications that involve dynamic loads in which the bidirectional and randomly directed fibre reinforcement is more convenient, as Köbler et al. [43] stated. In some arrangements, all fibre matting in composites can be arranged with polymer in one direction, as shown in Fig. 2. These layouts serve to illustrate the material's anisotropic behaviour. Many applications related to these fibre arrangements primarily utilize static loads.

In some circumstances, composites created for aerospace applications involving dynamic loads should behave like an isotropic material [45]. Structural integrity is essential in developing more efficient aircraft technologies for aerospace applications, specifically aircraft. This is because dynamic loading is dominant when the aircraft is in operation, especially in a critical regime [46]. For example, the fuselage must have high strength to sustain a highly complex loading without increasing the weight. For these circumstances, randomly oriented arrangements or layers are typically recommended. The relative positioning of the layers exhibits variation in orientation when the stacked layers are altered, as depicted in Fig. 3 [44]. The laminated materials were arranged in a stacked configuration, with each layer positioned at 45° relative to the adjacent layer. According to the experimental studies conducted by Pickering and Efendy [47] and Arun Prakash & Viswanathan [48], it has been observed that the inclusion of mat reinforcement in composites enables the material to effectively bear loads in several directions, owing to its specific orientation.

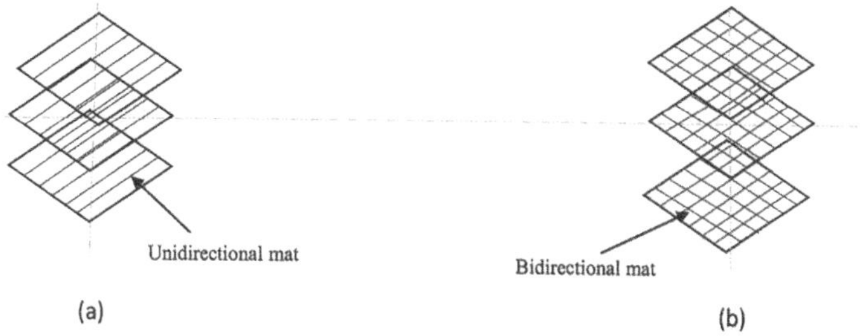

Fig. 2 Conventional geometry orientation of fibre in biocomposite application (**a**) unidirectional; (**b**) bidirectional [44]

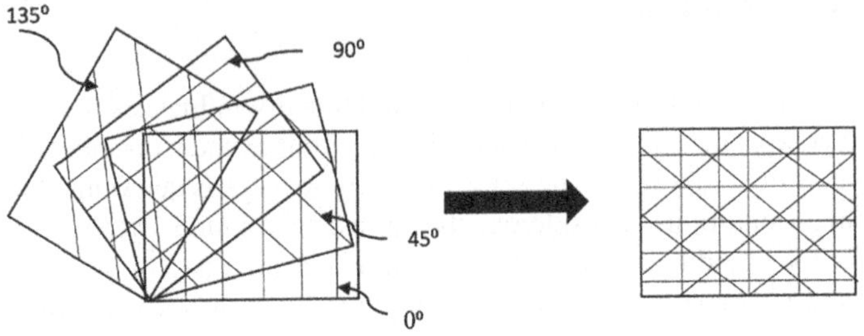

Fig. 3 Angled configurations of stacked matting [44]

Based on Fig. 3, the load applied to the mat orientations is distributed amongst the fibres in the warp and weft directions. According to the findings by Lotfi et al. [49], the load distribution on a mat reduces localized stress concentration compared to the case where the load is not distributed evenly. When used in composite materials, fibre matting arranged in this manner provides better load-bearing capacities in all directions. The arrangement of randomly oriented mat-reinforced composites under loading permits equal load distribution in all directions, and the material can withstand more significant applied stresses [49, 50]. Differences in fibre orientation also drive variations in the toughness of composite materials. Composites have less tensile strength if reinforced fibres are not perfectly perpendicular to the fracture planes, whereas fibres parallel to the fracture surface have less tensile strength [51]. Cording et al. [52] investigated the static and dynamic properties of propylene and lyocell cellulosic fibres. The study focused on analysing these features at different fibre angles of reinforcement, specifically 0°, 22.5°, 45°, 67.5°, and 90°. According to the authors' findings, the composites reinforced with a 0° angle exhibited superior tensile properties than those reinforced with different fibre angles. In contrast to a range of composites reinforced with different fibre angles, it was shown that the orientation of lyocell fibres at a 45° angle exhibited superior dynamic characteristics. When the oriented angle of unidirectional mat fibres is 0° with respect to the loading direction, composites demonstrate higher tensile strength. The tensile properties exhibited a decline when the orientation of the unidirectional fibre mat was increased to 90° relative to the direction of loading.

The flexural properties of composites are influenced by the arrangement of fibres, like how tensile strength is affected by fibre orientation. Typically, flexural testing is conducted to evaluate the bending capability of a material. According to Selmy et al. [53], the flexural strength of thermoplastic composites reinforced with unidirectional fibres is comparatively lower when compared to composites reinforced with bidirectional mats. The interlocking between fibres in bidirectional reinforcement occurs at many interlocking sites, providing flexural load resistance [54]. Figure 4 shows that interlocking is typically absent or negligible in other geometrically distinct mats, such as unidirectional and randomly oriented mats.

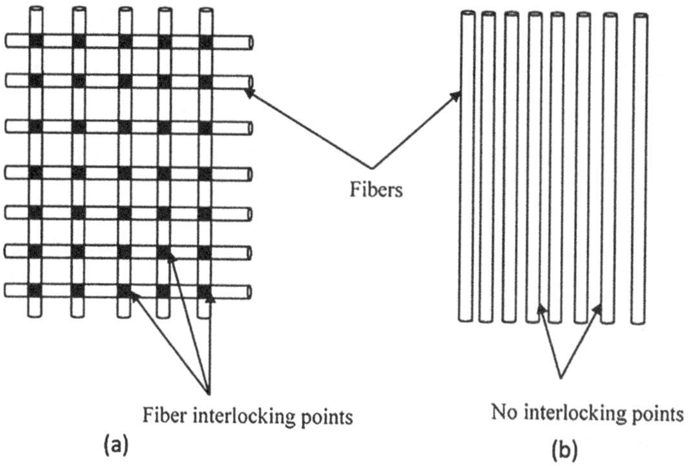

Fig. 4 Interlocking points in fibre orientation of (**a**) bidirectional and (**b**) unidirectional mat [44]

3.1 Hybrid Biocomposite Structure

The concept of hybridization is widely used in many fields, including mechanics, polymers, and chemical chemistry, as well as in metallurgy, physics, science and technology, and energy sources. Regardless of the study, the primary objective of this process is to combine three or more substances to achieve superior performance in a particular region instead of relying on each substance's capabilities for the intended application [55]. For a suggested application, composite performance or characteristics are compared to those of individual substances [56]. The disadvantages of composite materials can be minimized by integrating two or more fillers or fibres into a matrix material, which also allows replacing one reinforcement type with another [57]. One example of how the limitations of a particular type of reinforcement can be alleviated is through hybridization. For instance, combining natural fibres with synthetic fibres through hybridization reduces the inherent advantages of natural fibres, such as their biodegradability, ease of handling and processing, and utilization of renewable resources. However, this hybridization technique can decrease costs, weight, and environmental impacts associated with the production of composite materials [58]. Figure 5 depicts some of the possible configurations of hybrid fibre-reinforced biocomposites.

Several studies have been undertaken on developing hybrid biocomposites; however, these efforts have been limited to the fabrication and testing stages [60, 61]. There needs to be more information and reported data about the production and specific application of hybrid biocomposites in the aerospace industry. The closest information found was on the fabrication method of composites for aircraft parts that are also adaptable for natural fibres. This includes hand lay-up method to fabricate aircraft seats, non-structural parts, and interior cabin parts, vacuum infusion to fabricate vehicle floor interior cabin part, and compression moulding for nacelle

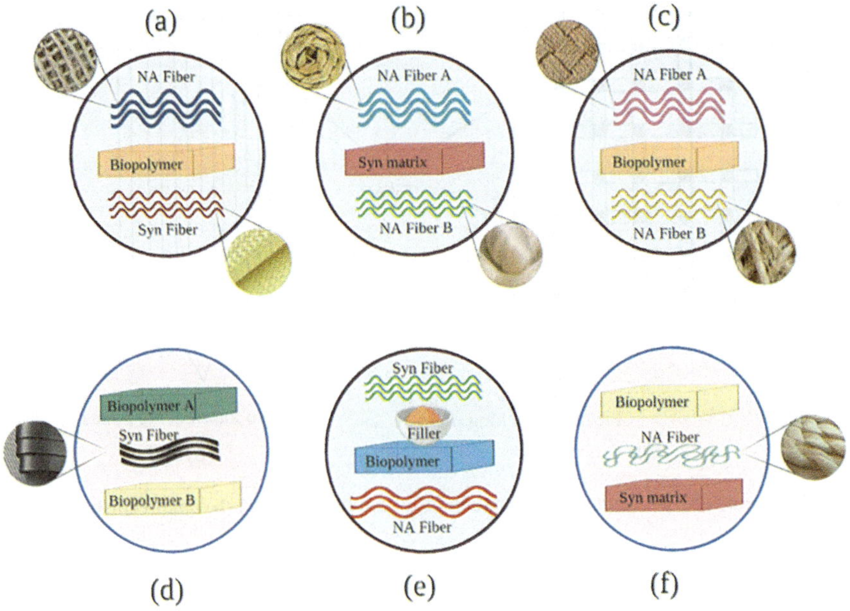

Fig. 5 Some configurations of hybrid biocomposites (NA: natural, Syn: synthetic): (**a**) hybridization of natural and synthetic fibres with biopolymer matrix; (**b**) hybridization of two or more natural fibres with synthetic matrix; (**c**) hybridization of two or more natural fibres with biopolymer matrix; (**d**) hybridization of two or more biopolymer matrices with synthetic fibre; (**e**) hybridization of natural and synthetic fibres with fillers with biopolymer matrix; (**f**) hybridization of biopolymer and synthetic matrices with natural fibre [59]

structure [38]. In relation to their potential application in the aerospace industry, the hybrid biocomposites reported are hybrid kenaf and glass composite for radome structure [62], and the use of natural fibre such as wood and coconut coir for making anechoic chamber [63, 64]. Despite that, more information on the production and specific application of hybrid biocomposites is needed. This step serves to narrow the gap and brings us closer to realizing the potential substitution of synthetic composites with biocomposites.

The mechanical properties of hybrid biocomposites can be affected by various factors, such as the mechanical characteristics of the matrix and reinforcements, the dispersion and distribution of the reinforcements, their volume fractions, aspect ratios, and interfacial adhesion with the polymer, the loading and testing conditions, the dimensions and orientation of the fibres, and surface modifications [60]. Figure 6 displays the tensile strength and tensile modulus of epoxy-based hybrid biocomposites while Fig. 7 displays the flexural strength and flexural modulus of epoxy-based hybrid biocomposites. For both tensile and flexural properties, glass, carbon, and Kevlar fibres are commonly used as part of the hybrid reinforcing fibres in biocomposites. Therefore, the tensile and flexural properties shown here are grouped with respect to the synthetic fibres. For example, in Fig. 6, it was found that hybrid sugar

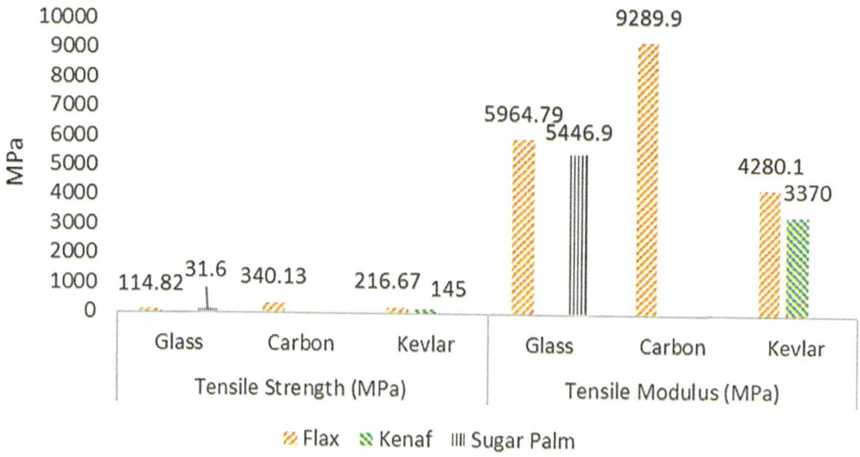

Fig. 6 Tensile strength and tensile modulus of epoxy-based hybrid biocomposites [65–67]

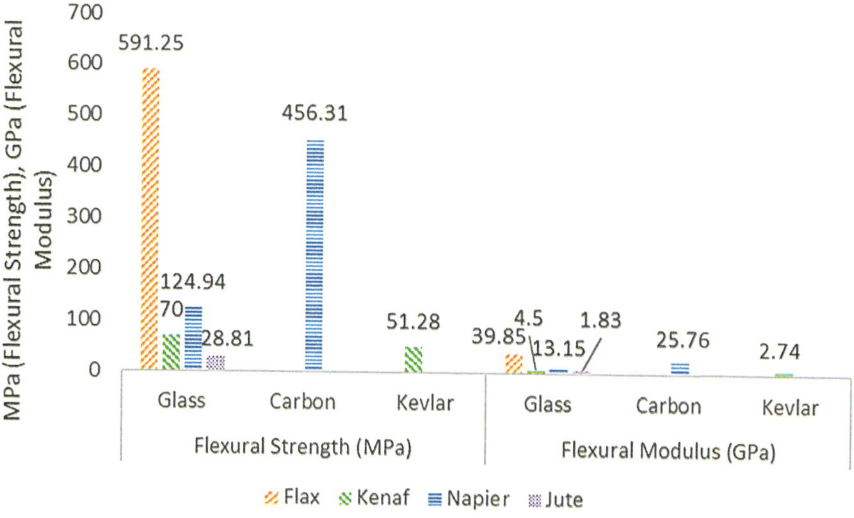

Fig. 7 Flexural strength and flexural modulus of epoxy-based hybrid biocomposites [66, 68–71]

palm/glass epoxy biocomposite produced the lowest tensile strength of 31.6 MPa while hybrid flax/carbon epoxy biocomposite exhibited the highest tensile strength with 340.13 MPa. Based on Figs. 6 and 7, it was found that the hybridization of natural fibre flax with synthetic fibre glass, carbon, and Kevlar often resulted in superior tensile strength, tensile modulus, flexural strength, and flexural modulus.

Figure 8 (a) displays the tensile strength and (b) tensile modulus of PLA based hybrid biocomposites while Fig. 8 (c) displays the flexural strength and (d) flexural modulus of Polypropylene based hybrid biocomposites. From Fig. 8a, a PLA based hybrid biocomposite of kenaf, bamboo, and coir showed relatively high tensile

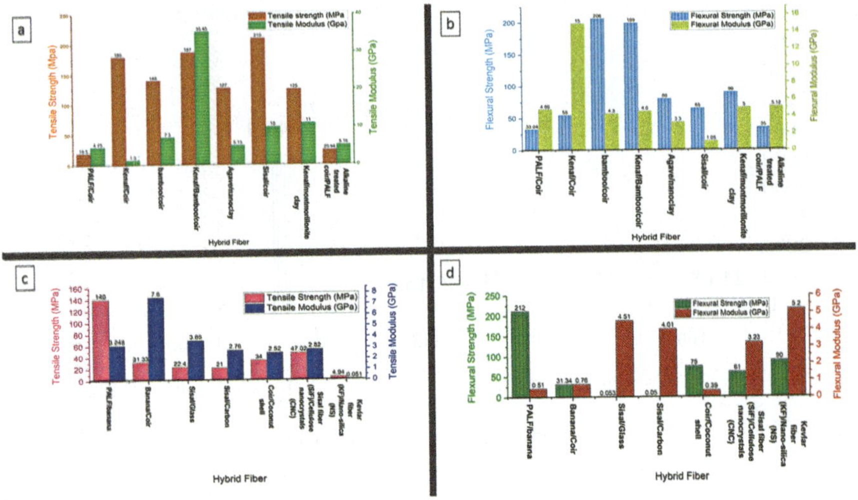

Fig. 8 (**a**)Tensile strength; (**b**) tensile modulus of PLA-based hybrid biocomposites; (**c**) tensile strength; (**d**) tensile modulus of polypropylene-based hybrid biocomposites [57, 71–82]

strength and tensile modulus. In addition to that, Fig. 8b shows that the same biocomposite displays high flexural strength of 199GPa. However, its flexural strength is 70% lower than PLA based hybrid kenaf/coir biocomposite.

In Fig. 8c, d, polypropylene based PALF/Banana hybrid composite displays the highest tensile strength and tensile modulus among all the hybrid biocomposites, indicating its superiority. On the other hand, it is worth noting that polypropylene based Banana/Coir hybrid biocomposite in Fig. 8c exhibits the highest tensile modulus of 7.6GPa and polypropylene based Kevlar/Nanosilica hybrid biocomposite has the highest flexural modulus of 5.2GPa.

3.2 *Hybrid Biocomposite in Aerospace Applications*

Due to the growing awareness of environmental concerns, natural fibre composites have been used as effective reinforcements in polymer matrix composites [83]. However, composites composed of identical reinforcing material systems exhibit poor performance because of the diverse loading conditions they experience over their service life. Consequently, many products that were once manufactured using conventional materials are presently being produced by employing hybrid composites to address this predicament. Hybrid composites find substantial application within the aerospace sector. The aerospace sector's relentless pursuit of enhancing civilian and military aircraft has resulted in high-performance structural materials. Among these materials, composite material has gained significant prominence and is now widely employed in producing aerospace components [84].

According to Samir et al. [85], they are perceived as more attractive due to their superior physical characteristics, including higher ratios of strength and stiffness to density. The utilization of modern aircraft structures has been enhanced by the emergence of robust reinforcements such as glass and carbon fibres, as well as developments in polymers [86, 87]. Besides, hybrid composites are employed in various aerospace applications, including missiles, big and small civil transport aircraft, helicopters, satellites, and military fighter aircraft. For example, George et al. [72] use a hybrid composite of ethylene propylene diene monomer rubber/Kevlar fibre based for rocket motor insulation materials and Zin et al. [38] revealed the uses of interior aircraft cabin using hybrid composite to replace the conventional materials.

4 Biocomposite Characteristics on Aircraft Structure

4.1 Radome Structure

Radome is an important structural component in an aircraft. It protects the antennas in the aircraft's nose section from the external environment, airflow, hail, lightning strikes, and bird strikes [88]. It is required to retain the electromagnetic performance of the antenna, thus maintaining the performance of the aircraft's radio guidance system [89]. In its manufacturing process, the materials must fulfil several characteristics to ensure they can be transparent to radio waves, which have a low dielectric constant and high toughness. The materials should also be able to minimize the attenuation of signals transmitted or received by the antenna. Significant differences in shape, size, and materials can be seen in radome manufacturing for commercial planes and fighter jets as they serve different purposes and missions [90].

Between 1970 and 1980, the radome was among the earliest parts made of polymer composites for the A300-B2 model aircraft [88]. The common fibre used for these polymer composites is the E-glass fibre reinforced in epoxy or polyester. Considering that natural fibre composites have been shown to have comparable properties to glass fibre composites in many other applications, research on the possibility of biocomposites as radome structures has been initiated [22]. Besides having a low density and high strength-to-weight ratio, natural fibre offers a low dielectric constant range from 2.8 to 3.2 compared to E-glass woven with epoxy and polyester, which ranges from 4.03 to 4.43 [91]. The low dielectric constant is one of the characteristics in the selection of material for radome.

During service, the radome structure is prone to low-velocity impact damage. This damage indirectly reduces its structural integrity and affects its dynamic characteristics, thus leading to excessive vibration from resonance. Chok et al. [92] investigated the dynamic properties of kenaf. The authors found that using kenaf in composites, either on its own or hybrid with glass composites, can stall the damage progression and reduce the effects of damage on the radome structure. Kenaf

Table 3 Mechanical properties of natural fibres and E-glass fibre [22]

Fibre	Tensile strength (MPa)	Tensile modulus (GPa)
E-glass	2000–3500	70
Kenaf	930	53
Bamboo	503	35.9
Banana	500	12
Pineapple leaf	170	6.21
Oil palm empty fruit bunch	0.1–0.4	1–9

composites showed the lowest natural frequency reduction during impact, which is 1–18% from its original value, followed by the kenaf/glass hybrid composites [92].

Preliminary comparisons between five different natural fibres were conducted to identify the most preferable natural fibre for the radome structure. The analysis suggested kenaf fibre as the highest potential natural fibre for radome structure due to its low dielectric constant and comparable stiffness to glass fibre. As an introduction, kenaf was suggested to be a hybrid with glass fibre to enhance its performance and improve its resistance to environmental conditions [22]. Table 3 compares the mechanical properties of natural fibres and glass fibre.

Radome is highly susceptible to impact from rain, freeze-thaw cycles, lightning strikes, and static electricity [93]. It must withstand delamination, fracture, and degradation as stated by the MIL-STD-7705B standard during those conditions. Therefore, besides fibre's mechanical properties, biocomposites' low-velocity impact properties are also important to be evaluated for radome application [94]. In an experimental analysis, kenaf/epoxy composites showed low resistance to impact damage at impact energy ranges from 3 to 9 J. Thus, hybrid kenaf/glass composites were suggested to enhance its load-carrying capacity. In both composites, the woven kenaf was seen to act as a stopping mechanism to the propagation of the matrix cracking and limit the propagation along the warp and weft directions as opposed to radial crack propagation [95].

4.2 *Wing Structure*

In a wing structure of a fixed-wing aircraft, the primary load-carrying structure, which acts as the structural centre of the wings, is known as the wing box. It becomes the attachment point for other wing components, such as leading-edge flaps, trailing-edge flaps, and wing-tip devices. To be among the main structural components of an aircraft, the strength properties of a wing box are the most important aspect to be fulfilled. Over the years, sustainability of the material and weight reduction have been emerging issues to be highlighted [96].

A detailed theoretical study has been conducted regarding a civil aircraft model, Airbus A320–200. The wing box made up from Ramie fibre saved 12–14% weight compared to 7000-series aluminium alloy due to its low density without compromising structural integrity. This study strongly believed that bio-based materials must be seriously given a space to be applied in structural applications, considering the integration with the structure's design, which needs to be further analysed [97].

5 Life Cycle Assessment (LCA) of Sustainability in Biocomposite Materials

To evaluate the sustainability of biocomposites and their environmental implications, a life cycle assessment methodology can be employed. This methodology entails the examination of the environmental implications associated with biocomposites throughout many stages of their life cycles, encompassing the extraction of raw materials and concluding with their disposal. Using a life cycle assessment (LCA) methodology makes it possible to determine and quantify the environmental implications associated with the manufacturing, utilization, and disposal of biocomposites. Given the critical role that natural fibres play in the life cycle of biocomposites, this assessment technique is crucial. The utilization of the life cycle assessment methodology in the context of biocomposites is crucial to establish the presence of suitable systems for treating these materials at the end of their life cycle. Furthermore, research on biocomposite assessment methods has identified knowledge and information gaps that must be addressed to effectively implement these procedures [3].

Life cycle assessment is a prominent tool for evaluating biocomposites' sustainability and environmental impact. Potential environmental impacts from material inputs and discharges can be evaluated by assessing the environmental aspects and impacts throughout the entire life cycle of biocomposites, including the gathering of energy and material inputs. The life cycle assessment results can then be interpreted to assist policymakers in making informed decisions regarding the environmental impact of biocomposites. Additionally, life cycle assessment can be used to compare the environmental impact of different products or systems, allowing for more sustainable choices. The life cycle assessment of biobased composites has shown several advantages over traditional petro-based composites in terms of energy use and environmental impacts [98]. For example, studies have shown that bio-based biocomposites have lower greenhouse gas emissions and embodied energy than conventional composite materials. These findings make biofibre-based biocomposites an attractive alternative material for aircraft manufacturers seeking to reduce their environmental footprint. However, it is essential to note that the steady supply of bio-based fibres can be challenging. Therefore, further research and development are needed to ensure the sustainable sourcing and production of natural fibres for biocomposites. In conclusion, life cycle assessment is a crucial tool for evaluating biocomposites' sustainability and environmental impact [3].

5.1 Advancements and Future Directions

Advancements in life cycle assessment methodologies can further enhance the accuracy and reliability of assessing the sustainability of biocomposites. These advancements can include developing more comprehensive and standardized data collection methods, incorporating social and economic factors into the assessment, and considering uncertainty and variability in the assessment results. Furthermore, future research should also focus on developing sustainable sourcing and production methods for natural fibres used in biocomposites. This could involve exploring more efficient farming practices, implementing recycling and waste management systems, and finding ways to minimize the environmental impact of processing and manufacturing. By addressing these challenges and implementing sustainable practices throughout the life cycle of biocomposites, their environmental impact can be further reduced. In addition, it is essential to consider the end-of-life treatment systems for biocomposites. This includes examining potential recycling or disposal methods to minimize environmental impact. Moreover, as the demand for biocomposites continues to grow, it is important to explore and develop new fibres, processes, and fillers that can improve the strength and versatility of biocomposites without compromising their sustainability. By doing so, biocomposites can be used in a broader range of applications and replace synthetic materials in various industries.

5.2 Challenges and Recommendations

Biocomposites offer the potential for sustainable alternatives to traditional materials in various industries. However, it is essential to assess the life cycle of biocomposites to ensure their sustainability and minimize environmental impact. Life cycle assessment techniques can provide valuable insights into the environmental implications of biocomposites at different stages of their life cycles. This assessment can help identify improvement areas in data collection methods, inclusion of social and economic factors, and consideration of uncertainty and variability. Furthermore, addressing challenges related to sustainable sourcing and production methods for natural fibres used in biocomposites is crucial. This can involve exploring more efficient farming practices, implementing recycling and waste management systems, and finding ways to minimize the environmental impact of processing and manufacturing. Furthermore, addressing challenges related to sustainable sourcing and production methods for natural fibres used in biocomposites is crucial. Several published works have reported on the same issue [99–101]. This can involve exploring more efficient farming practices, implementing recycling and waste management systems, and finding ways to minimize the environmental impact of processing and manufacturing. In addition, it is important to promote research and development efforts towards finding alternative materials that can replace natural fibres in biocomposites [102]. This can help reduce the demand for these resources and alleviate the pressure on sustainable sourcing [103]. Additionally, collaboration between industries, governments, and environmental organizations is essential to establish regulations and standards that ensure responsible sourcing and production of natural fibres for biocomposites [104].

5.2.1 Challenges in Sustainable Sourcing

Limited availability of sustainable natural fibres on increasing demand for sustainable products: As more consumers become aware of the environmental impact of their purchasing decisions, there is a growing demand for products made from sustainable natural fibres [105]. This puts pressure on sourcing companies to find enough suppliers who can provide these materials in a sustainable manner.

Complex supply chains: Sustainable sourcing often involves multiple stages, from harvesting the natural fibres to processing and manufacturing the final product. Managing and monitoring these complex supply chains can be challenging, as it requires ensuring that each step meets sustainability criteria.

Another challenge in sustainable sourcing is the lack of standardized certification for verifying sustainable practices. This makes it difficult for sourcing companies to accurately assess and compare the sustainability credentials of different suppliers. Additionally, without a standardized certification, it can be challenging for consumers to identify and support products that are truly sourced sustainably.

High costs associated with sustainable sourcing can also pose a barrier for sourcing companies. Implementing sustainable practices often requires investments in new technologies, training, and certifications, which can increase production costs [106]. These added expenses may be passed on to consumers, making sustainable products more expensive and potentially limiting their accessibility to a wider market.

5.2.2 Challenges in Sustainable Production Methods

Energy-intensive production processes can also be a challenge in achieving sustainability goals. Many industries rely heavily on fossil fuels for their production, which contributes to greenhouse gas emissions and environmental degradation [107]. Transitioning to more energy-efficient methods or renewable energy sources can be costly and require significant infrastructure changes. However, investing in sustainable production methods can lead to long-term cost savings and environmental benefits [108].

Chemical usage and its impact on the environment are another important aspect to consider in achieving sustainability goals. Many industries use chemicals in their production processes, which can have negative effects on ecosystems and human health. Finding alternatives to harmful chemicals or implementing stricter regulations can help minimize these impacts and promote a more sustainable approach to production [109]. Additionally, educating consumers about the importance of choosing environmentally friendly products can also play a crucial role in reducing chemical usage and its environmental impact.

Waste management and disposal of byproducts are also important considerations in reducing the environmental impact of chemical usage [110]. Proper handling and disposal of chemical waste can prevent contamination of water sources and soil, further protecting ecosystems and human health. Implementing efficient waste management practices, such as recycling or proper treatment methods, can help minimize the release of harmful chemicals into the environment.

5.2.3 Potential Solutions for Sustainable Sourcing

Promoting research and development for alternative natural fibres that can be used as substitutes for synthetic fibres can help reduce the reliance on chemical-intensive processes in textile production [111]. Additionally, implementing sustainable farming practices, such as organic agriculture, can minimize the use of chemical fertilizers and pesticides in growing natural fibres, further reducing the environmental impact of sourcing materials for various industries.

Encouraging cultivation and harvesting practices that promote sustainability can also contribute to the conservation of biodiversity and soil health [112]. Moreover, supporting fair trade initiatives can ensure that farmers and workers involved in the production of natural fibres receive fair wages and are provided with safe working conditions, promoting social and economic sustainability as well.

Establishing industry-wide standards and certifications for sustainable sourcing can help ensure that the production of natural fibres is environmentally responsible and ethically sourced [113]. Additionally, investing in research and development of innovative technologies can lead to more sustainable and efficient production processes, reducing the overall environmental impact of the industry.

5.2.4 Potential Solutions for Sustainable Production Methods

Investing in renewable energy sources for production processes can significantly reduce the carbon footprint of the industry [114] This can involve implementing solar panels or wind turbines to power factories and reducing reliance on fossil fuels. Furthermore, adopting circular economy principles, such as recycling and reusing materials, can minimize waste [115] and promote a more sustainable approach to production.

Implementing green chemistry practices to reduce chemical usage and minimize the release of harmful pollutants into the environment is another effective method. This can include using safer and more sustainable alternatives to traditional chemicals, as well as optimizing processes to reduce chemical waste. Additionally, implementing lean manufacturing techniques can help streamline production processes, reduce energy consumption, and minimize overall environmental impact [116].

Adopting efficient waste management and recycling strategies can also contribute to reducing environmental pollution. This can involve implementing proper segregation and disposal methods for different types of waste, as well as promoting recycling and reusing materials whenever possible [117]. By effectively managing waste, industries can minimize their contribution to landfills and reduce the need for raw material extraction, ultimately leading to a more sustainable and eco-friendly approach to production.

By addressing these challenges and implementing sustainable practices throughout the life cycle of natural fibre source, their environmental impact can be further reduced. In addition to addressing the environmental impact of biocomposites during their use and manufacturing, it is equally important to consider their end-of-life treatment systems.

6 Conclusion

This chapter provides an overview of a comprehensive review of research that examines recent investigations on the possible applications of biocomposite materials in the aerospace industry. These materials have been widely utilized in aircraft's interior and secondary components, demonstrating substantial practicality and effectiveness. In addition to the inherent uncertainties associated with natural fibre architectures, numerous other variables influence the performance of biocomposite materials. These aspects encompass the treatment of fibres, the incorporation of diverse fibres or fillers through hybridization, the orientation of fibres, and various other considerations. Extensive research has been undertaken to investigate the characteristics of design biocomposite materials. Considering the rigorous regulations imposed by the Federal Aviation Administration (FAA) regarding aircraft materials, further investigation is required to ascertain the viability of incorporating biocomposite materials into aircraft. Despite the aviation industry's efforts to adopt green materials due to the sustainability of raw materials, more research still needs to be conducted to realize the potential of biocomposite materials. Therefore, further investigation is required to develop biocomposites that possess comparable or superior qualities to conventional synthetic fibre-reinforced polymer composites often employed in aeroplanes, particularly the aircraft seats and nacelle structures.

Acknowledgement The authors would like to thank Universiti Putra Malaysia (UPM) for providing the support and resources necessary to conduct this review.

References

1. Lukowsky, D., & Gohla, A. (2022). Material failure analysis of three in-flight wing fractures of wooden aircraft. *Journal of Failure Analysis and Prevention, 22*(3), 1075–1084.
2. Kumar, S. (2023). *Biocomposites market size is anticipated to reach at a CAGR of 14.8% during the forecast period 2022–2030.* https://www.linkedin.com/pulse/biocomposites-market-size-anticipated-reach-cagr-148-during-kumar. Accessed 2 Sept 2023.

3. Climate Change. (2007). Working group III: Mitigation of climate change. Executive summary.. http://www.ipcc.ch/publications_and_data/ar4/wg3/en/ch11s11-es.html. Accessed 2 Sept 2023.
4. Hassan, T. H., Sobaih, A. E. E., & Salem, A. E. (2021). Factors affecting the rate of fuel consumption in aircrafts. *Sustainability, 13*(14), 8066.
5. Gomez-Campos, A., Vialle, C., Rouilly, A., et al. (2021). Natural fibre polymer composites-A game changer for the aviation sector? *Journal of Cleaner Production, 286*, 124986.
6. Graver, B., Zhang, K., & Rutherford, D. (2019). Emissions from commercial aviation, 2018. In *International Council on Clean Transportation*. https://theicct.org/publication/co2-emissions-from-commercial-aviation-2018/. Accessed 13 Sept 2023.
7. Bachmann, J., Yi, X., Tserpes, K., et al. (2021). Towards a circular economy in the aviation sector using eco-composites for interior and secondary structures. Results and recommendations from the EU/China project ECO-COMPASS. *Aerospace, 8*(5), 131.
8. Mohan, S. V., & Katakojwala, R. (2021). The circular chemistry conceptual framework: A way forward to sustainability in industry 4.0. *Current Opinion in Green and Sustainable Chemistry, 28*, 100434.
9. Kobayashi, T., & Nakajima, L. (2021). Sustainable development goals for advanced materials provided by industrial wastes and biomass sources. *Current Opinion in Green and Sustainable Chemistry, 28*, 100439.
10. Katunin, A., Krukiewicz, K., Herega, A., & Catalanotti, G. (2016). Concept of a conducting composite material for lightning strike protection. *Advances in Materials Science, 16*(2), 32–46.
11. Meng, F., Cui, Y., Pickering, S., & McKechnie, J. (2020). From aviation to aviation: Environmental and financial viability of closed-loop recycling of carbon fibre composite. *Composites Part B: Engineering, 200*, 108362.
12. Zhang, J., Lin, G., Vaidya, U., et al. (2023). Past, present and future prospective of global carbon fibre composite developments and applications. *Composites Part B: Engineering, 110463*, 110463.
13. Helping Aircraft Decommisioning. (2023). https://www.iata.org/en/programs/environment/aircraft-decommissioning/. Accessed 2 Sept 2023.
14. Ramachandran, K., Gnanasagaran, C. L., & Vekariya, A. (2023). Life cycle assessment of carbon fibre and bio-fibre composites prepared via vacuum bagging technique. *Journal of Manufacturing Processes, 89*, 124–131.
15. El-Ghezal, S., & Khiari, B. (2023). Palm wastes for bio-based materials production. In M. Jeguirim, B. Khiari, & S. Jellali (Eds.), *Palm trees and fruits residues* (1st ed., pp. 191–222). Academic.
16. Hassan, A., Salema, A. A., Ani, F. N., et al. (2010). A review on oil palm empty fruit bunch fibre-reinforced polymer composite materials. *Polymer Composites, 31*(12), 2079–2101.
17. Rudin, A., & Choi, P. (2012). *The elements of polymer science and engineering*. Academic.
18. Christian, S. J. (2020). Natural fibre-reinforced noncementitious composites (biocomposites). In K. A. Harries & B. Sharma (Eds.), *Nonconventional and vernacular construction materials* (2nd ed., pp. 169–187). Woodhead Publishing.
19. Ramesh, M., & Deepa, C. (2021). Mechanical properties of natural and synthetic fibre reinforced hybrid composites. In A. Khan, S. M. Rangappa, S. Siengchin, et al. (Eds.), *Hybrid natural fibre composites* (1st ed., pp. 309–325). Woodhead Publishing.
20. Sionkowska, A. (2011). Current research on the blends of natural and synthetic polymers as new biomaterials. *Progress in Polymer Science, 36*(9), 1254–1276.
21. Santos, C. V. D., Leiva, D. R., Costa, F. R., et al. (2016). Materials selection for sustainable executive aircraft interiors. *Materials Research, 19*, 339–352.
22. Haris, M. Y., Laila, D., Zainudin, E. S., et al. (2011). Preliminary review of biocomposites materials for aircraft radome application. *Key Engineering Materials, 471*, 563–567.
23. Aircraft Materials Fire Test Handbook. (2023). *Federal Aviation Administration. Handbook: FAA Fire Safety*. Accessed 2 Sept 2023.

24. John, M. J., Anandjiwala, R. D., Wambua, P. et al. (2008). Bio-based structural composite materials for aerospace applications. In: *South African International Aerospace Symposium, Stellenbosch, South Africa*, 14–16 Sept 2008.
25. Boegler, O., Kling, U., Empl, D. et al. (2014). *Potential of sustainable materials in wing structural design*. In: Deutsche Gesellschaft für Luft-und Raumfahrt-Lilienthal-Oberth eV Bonn, Germany, 16–18 Sept 2014.
26. Niedermann, P., Szebényi, G., & Toldy, A. (2015). Characterization of high glass transition temperature sugar-based epoxy resin composites with jute and carbon fibre reinforcement. *Composites Science and Technology, 117*, 62–68.
27. GreeHenschel, K. A. (2019). *Biocomposites in aviation structures on the example of flax and its hybrids*. Dissertation, Technische Universität München.
28. Karunakaran, S., et al. (2016). Flammability of self-extinguishing kenaf/ABS nanoclays composite for aircraft secondary structure. In *IOP conference series: Materials science and engineering* (Vol. 152, p. 012068). IOP Publishing.
29. Martínez, X., Pons-Prats, J., Turon, F., et al. (2022). Multi-objective multi-scale optimization of composite structures, application to an aircraft overhead locker made with bio-composites. *Math, 165*, 1–21.
30. Rmanan, G., Akshatha, R. D., Manvi, A. U., et al. (2022). Investigation of bio degradable natural fibres reinforced hybrid composites for aircraft structures. *Materials Today: Proceedings, 52*, 1211.
31. Karthigairajan, M., Nagarajan, P., Rabiraja Malarvannan, R., et al. (2021). Effect of silane-treated rice husk derived biosilica on visco-elastic, thermal conductivity, and hydrophobicity behavior of epoxy biocomposite coating for air-duct application. *Si, 13*, 4421–4430.
32. da Costaa, R. R., Eloyb, F. D. S., de Medeirosc, R. et al. (2017). Experimental dynamic investigations of bio and synthetic composites for use in aircraft interior structures. In: *Conference: meeting on aeronautical composite materials and structures*.
33. Eloy, F. D. S., Costa, R. R., De Medeiros, R., et al. (2015). Comparison between mechanical properties of bio and synthetic composites for use in aircraft interior structures. In *Meeting on aeronautical composite materials and structures*. University of São Paulo.
34. Wu, T., Williams, G., Wang, Y., et al. (2023). Advanced fibre materials for controlled release, tissue repair, and regenerative medicine. *Front, 10*, 1244284.
35. Madueke, C. I., Mbah, O. M., & Umunakwe, R. (2023). A review on the limitations of natural fibres and natural fibre composites with emphasis on tensile strength using coir as a case study. *Polymer Bulletin, 80*(4), 3489–3506.
36. Kim, N. K., Dutta, S., & Bhatacharyya, D. (2018). A review of flammability of natural fibre reinforced polymeric composites. *Composites Science and Technology, 162*, 64–78.
37. Gopi, S., Balakrishnan, P., & Sreekala, M. S. (2017). Green materials for aerospace industries. In *Biocomposites for high-performance applications* (p. 307). Woodhead Publishing.
38. Zin, M. H., Razzi, M. F., Othman, S., et al. (2016). A review on the fabrication method of bio-sourced hybrid composites for aerospace and automotive applications. In *IOP conference series: Materials science and engineering* (Vol. 152, p. 012041). IOP Publishing.
39. De Leon, A. C., Chen, Q., Palaganas, N. B., et al. (2016). High performance polymer nanocomposites for additive manufacturing applications. *Reactive and Functional Polymers, 103*, 141–155.
40. Ghori, S. W., Siakeng, R., Rasheed, M., et al. (2018). The role of advanced polymer materials in aerospace. In *Sustainable composites for aerospace applications* (p. 19). Woodhead Publishing.
41. Dicker, M. P., Duckworth, P. F., Baker, A. B., et al. (2014). Green composites: A review of material attributes and complementary applications. *Composites, 56*, 280–289.
42. Ben Amor, I., Rekik, H., Kaddami, H., et al. (2010). Effect of palm tree fibre orientation on electrical properties of palm tree fibre-reinforced polyester composites. *Journal of Composite Materials, 44*(13), 1553–1568.

43. Köbler, J., Schneider, M., Ospald, F., et al. (2018). Fibre orientation interpolation for the multiscale analysis of short fibre reinforced composite parts. *Computational Mechanics, 61*, 729–750.
44. Manral, A., Bajpai, P. K., Ahmad, F., et al. (2021). Processing of sustainable thermoplastic based biocomposites: A comprehensive review on performance enhancement. *Journal of Cleaner Production, 316*, 128068.
45. Jaafar, J., Siregar, J. P., Mohd Salleh, S., et al. (2019). Important considerations in manufacturing of natural fibre composites: A review. *International Journal of Precision Engineering and Manufacturing-Green Technology, 6*, 647–664.
46. Zhou, J., Liu, J., Zhang, X., et al. (2019). Experimental and numerical investigation of high velocity soft impact loading on aircraft materials. *Aerospace Science and Technology, 90*, 44–58.
47. Pickering, K. L., & Efendy, M. A. (2016). Preparation and mechanical properties of novel bio-composite made of dynamically sheet formed discontinuous harakeke and hemp fibre mat reinforced PLA composites for structural applications. *Industrial Crops and Products, 84*, 139–150.
48. Arun Prakash, V. R., & Viswanathan, R. (2019). Fabrication and characterization of silanized echinoidea fillers and kenaf fibre-reinforced Azadirachta-indica blended epoxy multi-hybrid biocomposite. *International Journal of Plasticity, 23*, 207–217.
49. Lofti, A., Li, H., Dao, D. V., et al. (2021). Natural fibre–reinforced composites: A review on material, manufacturing, and machinability. *Journal of Thermoplastic Composite Materials, 33*(8), 1145–1171.
50. Mann, G. S., Singh, L. P., Kumar, P. (2020). Green composites: A review of processing technologies and recent applications. *J. Thermoplast, 33*(8): 1145–1171.
51. Abd-Ali, N. K., & Madeh, A. R. (2016). Effect of fibre orientation angles on mechanical behavior of car bumper composite. *Kufa Journal of Engineering, 7*(3), 27.
52. Cording, M., Bechtold, R., & Pham, T. (2018). Effect of fibre orientation on the mechanical properties of polypropylene–lyocell composites. *Cellulose, 25*, 7197–7210.
53. Selmy, A. I., Elsesi, A. R., Azab, N. A., et al. (2012). Interlaminar shear behavior of unidirectional glass fibre (U)/random glass fibre (R)/epoxy hybrid and non-hybrid composite laminates. *Composites Part B: Engineering, 43*(4), 1714–1719.
54. Naresh, K., Krishnapillai, S., & Velmurugan, R. (2017). Effect of fibre orientation on carbon/epoxy and glass/epoxy composites subjected to shear and bending. *Solid State Phenomena, 267*, 103–108.
55. Mochane, M. J., Mokhena, T. C., Mokhuthu, T. H., et al. (2019). Recent progress on natural fibre hybrid composites for advanced applications: A review. *Express Polymer Letters, 13*(2), 159–198.
56. Potluri R (2018) Natural fibre-based hybrid bio-composites: Processing, characterization, and applications. : . Springer , p 1–46.
57. Siakeng, R., Jawaid, M., Ariffin, H., et al. (2018). Thermal properties of coir and pineapple leaf fibre reinforced polylactic acid hybrid composites. In *IOP conference series: Materials science and engineering* (Vol. 368, p. 012019). IOP Publishing.
58. Asim, M., Jawaid, M., Paridah, M. T., et al. (2019). Dynamic and thermo-mechanical properties of hybridized kenaf/PALF reinforced phenolic composites. *Polymer Composites, 40*(10), 3814–3822.
59. Bahrami, M., Abenojar, J., & Martínez, M. Á. (2020). Recent progress in hybrid biocomposites: Mechanical properties, water absorption, and flame retardancy. *Maternité, 13*(22), 5145.
60. Guna, V., Ilangovan, M., Ananthaprasad, M. G., et al. (2018). Hybrid biocomposites. *Polymer Composites, 39*, 30–54.
61. Othman, F. E. C., & Hassan, S. A. (2023). Hybrid biocomposites—Combining synthetic nanofillers and natural fibres in composite structures. In *Synthetic and Natural Nanofillers in Polymer Composites* (pp. 259–272).

62. Zhou J, Ye F, Cheng L et (2023) Development of high-temperature wave-transparent nitride-based CFCMCs for aircraft radomes. Composites Part A: Applied Science and Manufacturing 107444, 167.
63. Ward, K. A., Winter, S. R., Cross, D. S., Robbins, J. M., Mehta, R., Doherty, S., & Rice, S. (2021). Safety systems, culture, and willingness to fly in autonomous air taxis: A multi-study and mediation analysis. *Journal of Air Transport Management, 91*(C), 101975.
64. Bhatnagar, A. (Ed.). (2016). *Lightweight ballistic composites: Military and law-enforcement applications*. Woodhead Publishing.
65. Ismail, K. I., Sultan, M. T. H. H., Shah, A. U., et al. (2018). Tensile properties of hybrid biocomposite reinforced epoxy modified with carbon nanotube (CNT). *Bioresources, 13*(1), 1787–1800.
66. Yahaya, R., Sapuan, S. M., Jawaid, M., et al. (2015). Effect of layering sequence and chemical treatment on the mechanical properties of woven kenaf–aramid hybrid laminated composites. *Materials and Design, 67*, 173–179.
67. Safri, S. N. A., Sultan, M. T. H., & Shah, A. U. M. (2020). Characterization of benzoyl treated sugar palm/glass fibre hybrid composites. *Journal of Materials Research and Technology, 9*, 11563–11573.
68. Ridzuan, M. J. M., Majid, M. A., & Khasri, A. (2019). Flexural properties of hybrid synthetic/Napier fibres reinforced epoxy composites. In *IOP conference series: Materials science and engineering* (Vol. 670, p. 012034). IOP Publishing.
69. Muhammad, Y. H., Ahmad, S., Abu Bakar, M. A., et al. (2015). Mechanical properties of hybrid glass/kenaf fibre-reinforced epoxy composite with matrix modification using liquid epoxidised natural rubber. *Journal of Reinforced Plastics and Composites, 34*, 896–906.
70. Braga, R. A., & Magalhaes, P. A., Jr. (2015). Analysis of the mechanical and thermal properties of jute and glass fibre as reinforcement epoxy hybrid composites. *Materials Science and Engineering: C, 56*, 269–273.
71. Manteghi, S., Mahboob, Z., Fawaz, Z., et al. (2017). Investigation of the mechanical properties and failure modes of hybrid natural fibre composites for potential bone fracture fixation plates. *Journal of the Mechanical Behavior of Biomedical Materials, 65*, 306–316.
72. George, K., Mohanty, S., Biswal, M., et al. (2021). Thermal insulation behaviour of Ethylene propylene diene monomer rubber/kevlar fibre based hybrid composites containing Nanosilica for solid rocket motor insulation. *Journal of Applied Polymer Science, 138*(9), 49934.
73. Yusoff, R. B., Takagi, H., & Nakagaito, A. N. (2016). Tensile and flexural properties of polylactic acid-based hybrid green composites reinforced by kenaf, bamboo and coir fibres. *Industrial Crops and Products, 94*, 562–573.
74. Del Campo, A. M., Robledo-Ortíz, J. R., Arellano, M., et al. (2020). GLYCIDYL methacrylate-compatibilized poly (lactic acid)/nanoclay/agave fibre hybrid biocomposites: Effect on the physical and mechanical properties. *Revista Mexicana de Ingenieria Quimica, 19*(1), 455–469.
75. Duan, J., Wu, H., Fu, W., et al. (2018). Mechanical properties of hybrid sisal/coir fibres reinforced polylactide biocomposites. *Polymer Composites, 39*, 188–199.
76. Ramesh, P., Prasad, B. D., & Narayana, K. L. (2019). Morphological and mechanical properties of treated kenaf fibre/MMT clay reinforced PLA hybrid biocomposites. In *AIP conference proceedings* (Vol. 2057). AIP Publishing.
77. Rahman, M., Das, S., & Hasan, M. (2018). Mechanical properties of chemically treated banana and pineapple leaf fibre reinforced hybrid polypropylene composites. *Advances in Materials and Processing Technologies, 4*(4), 527–537.
78. Bujjibabu, G., Das, V. C., Ramakrishna, M., et al. (2018). Mechanical and water absorption behavior of natural fibres reinforced polypropylene hybrid composites. *Maternité, 5*, 12249–12256.
79. Gupta, M. K., & Srivastava, R. K. (2016). Mechanical, thermal and water absorption properties of hybrid sisal/jute fibre reinforced polymer composite. *Indian Journal of Engineering and Materials Sciences, 23*, 231–238.

80. Aslan, M., Tufan, M., et al. (2018). Tribological and mechanical performance of sisal-filled waste carbon and glass fibre hybrid composites. *Composites Part B: Engineering, 40*, 241–249.
81. Yahaya, R., Sapuan, S. M., Jawaid, M., et al. (2016). Effect of fibre orientations on the mechanical properties of kenaf–aramid hybrid composites for spall-liner application. *Defence Technology, 12*, 52–58.
82. Agarwal, J., Mohanty, S., et al. (2021). Influence of cellulose nanocrystal/sisal fibre on the mechanical, thermal, and morphological performance of polypropylene hybrid composites. *Polymer Bulletin, 78*, 1609–1635.
83. Puttegowda, M., Rangappa, S. M., Jawaid, M., et al. (2018). Potential of natural/synthetic hybrid composites for aerospace applications. In *Sustainable composites for aerospace applications* (pp. 315–351).
84. Thyavihalli Girijappa, Y. G., Mavinkere Rangappa, S., Parameswaranpillai, J., et al. (2019). Natural fibres as sustainable and renewable resource for development of eco-friendly composites: A comprehensive review. *Frontiers in Materials, 6*, 26.
85. Samir, A., Ashour, F. H., Hakim, A., et al. (2022). Recent advances in biodegradable polymers for sustainable applications. *npj Materials Degradation, 6*(1), 68.
86. Awais, H., Nawab, Y., Amjad, A., et al. (2021). Environmental benign natural fibre reinforced thermoplastic composites: A review. *Composites Part C: Open Access, 4*, 100082.
87. Asyraf, M. R. M., Syamsir, A., Zahari, N. M., et al. (2022). Product development of natural fibre-composites for various applications: Design for sustainability. *Polymer Journal, 14*(5), 920.
88. Verma, D., Fortunati, E., & Jain, S. (Eds.). (2019). *Biomass, biopolymer-based materials, and bioenergy: Construction, biomedical, and other industrial applications.* Woodhead Publishing.
89. Zhou, J., Ye, F., Cheng, L., et al. (2023). Development of high-temperature wave-transparent nitride-based CFCMCs for aircraft radomes. *Composites Part A: Applied Science and Manufacturing, 167*, 107444.
90. Khatavkar, N., & Balasubramanian, K. (2016). Composite materials for supersonic aircraft radomes with ameliorated radio frequency transmission – A review. *RSC Advances, 6*(8), 6709–6718.
91. Shanmugasundram, H. P. P. V., Jayamani, E., & Soon, K. H. (2022). A comprehensive review on dielectric composites: Classification of dielectric composites. *Renewable and Sustainable Energy Reviews, 157*, 112075.
92. Chok, E. Y. L., Majid, D. L. A. A., & Harmin, M. Y. (2017). Effect of low velocity impact damage on the natural frequency of composite plates. *IOP Conference Series Materials Science and Engineering, 270*, 012025.
93. Lang, R. L. (1994). Radomes having vinyl foam core construction. Washington, US.
94. Jamal, Q. M. (2016). *Validation of low velocity impact on biocomposite flat plate laminates.* Master thesis, Universiti Putra Malaysia.
95. Majid, D. L., Jamal, Q. M., & Manan, N. H. (2018). Low-velocity impact performance of glass fibre, kenaf fibre, and hybrid glass/kenaf fibre reinforced epoxy composite laminates. *BioResources, 13*(4), 8839–8852.
96. Locatelli, D., Mulani, S. B., & Kapania, R. K. (2011). Wing-box weight optimization using curvilinear spars and ribs (SpaRibs). *Journal of Aircraft, 48*(5), 1671–1684.
97. Boegler, O., Kling, U., Empl, D., et al. (2015). Potential of sustainable materials in wing structural design. In *Deutscher Luft- und Raumfahrtkongress* (pp. 16–18).
98. La Rosa, A. D., Recca, G., Summerscales, J., Latteri, A., Cozzo, G., & Cicala, G. (2014). Bio-based versus traditional polymer composites. A life cycle assessment perspective. *Journal of Cleaner Production, 74*, 135–144.
99. Andrew, J. J., & Dhakal, H. N. (2022). Sustainable biobased composites for advanced applications: recent trends and future opportunities–A critical review. *Composites Part C: Open Access, 7*, 100220.

100. Akter, M., Uddin, M. H., & Tania, I. S. (2022). Biocomposites based on natural fibers and polymers: A review on properties and potential applications. *Journal of Reinforced Plastics and Composites, 41*(17–18), 705–742.
101. Vilaplana, F., Strömberg, E., & Karlsson, S. (2010). Environmental and resource aspects of sustainable biocomposites. *Polymer Degradation and Stability, 95*(11), 2147–2161.
102. Rani, A. M., Yuliawati, T., & Agustiningsih, D. D. (2019, March). How to Change MSME's Paradigm in Applying Green Industry Principles? In *Social and humaniora research symposium (SoRes 2018)* (pp. 236–239). Atlantis Press.
103. Mohammed, A. A., Hasan, Z., Omran, A. A. B., Kumar, V. V., Elfaghi, A. M., Ilyas, R. A., & Sapuan, S. M. (2022). Corn: Its structure, polymer, fibre, composite, properties, and applications. *Polymers, 14*(20), 4396.
104. Gairola, S. P., Tyagi, Y., Gangil, B., & Jha, K. (2021). *Physio-mechanical & wear performance of banana fibre/walnut powder based epoxy composites* (p. 42). Acta Innovations.
105. Vieira, L. M. G., Dobah, Y., dos Santos, J. C., Panzera, T. H., Campos Rubio, J. C., & Scarpa, F. (2022). Impact properties of novel natural fibre metal laminated composite materials. *Applied Sciences, 12*(4), 1869.
106. Freze, T., & Nurova, O. (2021). Green brands and sustainable entrepreneurship. In *E3S web of conferences* (Vol. 250, p. 04007). EDP Sciences.
107. Halkos, G. E., & Gkampoura, E. C. (2020). Reviewing usage, potentials, and limitations of renewable energy sources. *Energies, 13*(11), 2906.
108. Sari, S., & Setiyono, J. (2022). The development of green investment and its policies in the regulation of the Indonesian Government. *Lambung Mangkurat Law Journal, 7*(2), 118–131.
109. Thiam, B. G., El Magri, A., Vanaei, H. R., & Vaudreuil, S. (2022). 3D printed and conventional membranes—A review. *Polymers, 14*, 1023. Polymers and Their Application in 3D Printing, 159.
110. Nawaz, A., Chen, J., Su, X., & Zahid Hassan, H. M. (2022). Material based penalty-cost quantification model for construction projects influencing waste management. *Frontiers in Environmental Science, 10*, 807359.
111. Sarkar, F., Akonda, M., & Shah, D. U. (2020). Mechanical properties of flax tape-reinforced thermoset composites. *Materials, 13*(23), 5485.
112. Kraska, P., Andruszczak, S., Gierasimiuk, P., & Rusecki, H. (2021). The effect of subsurface placement of mineral fertilizer on some soil properties under reduced tillage soybean cultivation. *Agronomy, 11*(5), 859.
113. Melchor-Martínez, E. M., Macías-Garbett, R., Alvarado-Ramírez, L., Araújo, R. G., Sosa-Hernández, J. E., Ramírez-Gamboa, D., et al. (2022). Towards a circular economy of plastics: An evaluation of the systematic transition to a new generation of bioplastics. *Polymers, 14*(6), 1203.
114. De Meester, B., Vanholme, R., Mota, T., & Boerjan, W. (2022). Lignin engineering in forest trees: From gene discovery to field trials. *Plant Communications, 3*, 100465.
115. Wallner, T. S., Magnier, L., & Mugge, R. (2020). An exploration of the value of timeless design styles for the consumer acceptance of refurbished products. *Sustainability, 12*(3), 1213.
116. de Souza, A. J. S., Luciano, É. L., Fioravante, I. A., Rosa, J. L., de Melo, R. H., & Ribeiro, R. B. (2022). Management allied to industry: Development of metallic packaging "Rack" applied to the automotive sector. *International Journal of Business Administration, 13*(2), 1.
117. Söküt Açar, T., & Ayman Öz, N. (2020). The determination of optimal cluster number by Silhouette index at clustering of the European Union member countries and candidate Turkey by waste indicators. *Pamukkale Üniversitesi Mühendislik Bilimleri Dergisi, 26*(3), 481–487.

Additive Manufacturing Towards Sustainable Aerospace Structures

Joshua Rodrigues, Simon Barter, and Raj Das

Nomenclature

AM	Additive Manufacturing
CAD	Computer-Aided Design
DED	Directed Energy Deposition
DFAM	Design For Additive Manufacturing
FEM	Finite Element Method
ISRU	In Situ Resource Utilisation
ISS	International Space Station
PBF	Powder Bed Fusion

1 Introduction

The continual growth of aerospace operations and an interest in increasing the frequency of space launches have led to the development of innovative structures and materials that aim to minimise the associated environmental impact of these activities. The aerospace industry is one of the primary contributors to environmental pollution in both air and noise, thus epitomising the need for more sustainable development. Unprecedented growth in the aerospace industry has led to a surge in operations that continue to add to these environmental effects, prompting the need to address these concerns [28]. Several methods have already been implemented that aim to reduce the environmental impact of the aerospace industry, including the use of more ecological fuel sources, designing optimal trajectory paths and new policy implementation [29]. These efforts provide a proactive step towards creating a more environmentally conscious industry moving forward. Further initiatives are

J. Rodrigues (✉) · S. Barter · R. Das
School of Engineering, RMIT University, Melbourne, Australia
e-mail: joshua.rodrigues@student.rmit.edu.au

T. H. Karakoc et al. (eds.), *Sustainable Materials and Manufacturing Techniques in Aviation*, Sustainable Aviation, https://doi.org/10.1007/978-3-031-62987-7_3

still required across several aerospace fields to achieve a more sustainable outlook, including structures and materials, aerodynamics and propulsion systems.

In terms of aerospace structures, the demand for more lightweight and efficient components to address the rise in environmental concerns has prompted significant advancements in innovative structures and materials technologies. In particular, additive manufacturing (AM) is a leading example of such technological novelty that has revolutionised the way structures are designed and created. Since its initial use for rapid prototyping, AM technology has matured to become an established fabrication method across several industries including the aerospace, automotive and biomedical sectors [2]. In AM processes, components are often fabricated layer-by-layer, which presents numerous manufacturing advantages compared to traditional counterparts in subtractive processes which includes expanding the design freedom achievable and the opportunity to explore novel structural concepts [13]. These properties make AM particularly appealing to the aerospace industry and have since become adopted as a practised manufacturing method. The benefits of AM have boarded a new era for aerospace sustainability to produce structures that are not only less damaging to the environment but also offer a more efficient production method [30].

The role of AM technology as a pathway towards creating more sustainable and efficient structures is presented in this chapter. An overview of AM technology in aerospace is introduced to outline the uniqueness of AM processes and its close link to perform structural optimisation. The benefits of these characteristics towards a more sustainable sector are then discussed, signifying the advantages of weight reduction, reduced material processing, decentralised manufacturing and production flexibility. The current challenges and limitations are also presented in this chapter which highlights the present difficulties in the widescale use of AM in the aerospace industry, including quality assurance and production limitations. Stemming from this, the future opportunities for AM in the aerospace sector are presented to signify the potential growth of this technology. This includes the opportunities for the integration of AM at various life cycle stages, the use of in-space manufacturing capabilities and the prospect of in situ resource utilisation (ISRU). The culmination of this aims to demonstrate the potential of AM technology as a viable method to contribute towards more sustainable and efficient aerospace structures and an overall sector.

2 Additive Manufacturing in Aerospace

2.1 *Additive Manufacturing Technology*

As opposed to traditional, subtractive manufacturing techniques that often remove material from a bulk source, AM methods add layers sequentially to build a structure. Several classes of AM exist which can be applied to numerous material categories including polymers, ceramics and metals [33]. The process of AM can generally be understood as two main actions; the first being material deposition, followed by material solidification [6]. Typically, the material comes in the form of a powder or wire feedstock, that is then deposited onto a substrate surface where it solidifies. Successive layers are then applied in the same manner to create the final component geometry. Several methods of AM exist, with powder bed fusion (PBF) and directed energy deposition (DED) being commonly used in the aerospace industry [4].

PBF techniques can utilise either a laser or electron beam to melt and solidify the provided feedstock. In this process, a feedstock in powder form is distributed over a build plate where it is selectively sintered and solidified. The build plate then lowers, and a new layer of feedstock is distributed and sintered. This process is then repeated to eventually create the 3-dimensional part. PBF is known to produce components with high geometric precision, which is advantageous for aerospace applications. A schematic of this process using a laser is shown in Fig. 1a. Dissimilar to PBF, in DED processes the material feedstock is deposited locally into the meltpool created by an energy source, such as a laser or electron beam. In this process, the feedstock can be in the form of a powder similar to PBF or as a wire filament, both depicted in Fig. 1b. As this technique does not require a powder bed, it can offer greater geometrical freedom and accommodate larger build sizes which make it beneficial for larger components.

The generative nature of AM processes provides several unique characteristics that are advantageous to many industries. Particularly from an application perspective, AM enables the ability to significantly enhance the design freedom of structures by removing the constraints often imposed by traditional manufacturing processes [9]. Furthermore, this technology provides unprecedented control over the shape, composition and function during the design of structures, allowing for complex and internal features to be implemented [13]. These traits provide several unique opportunities to increase the effectiveness of AM technology for aerospace applications.

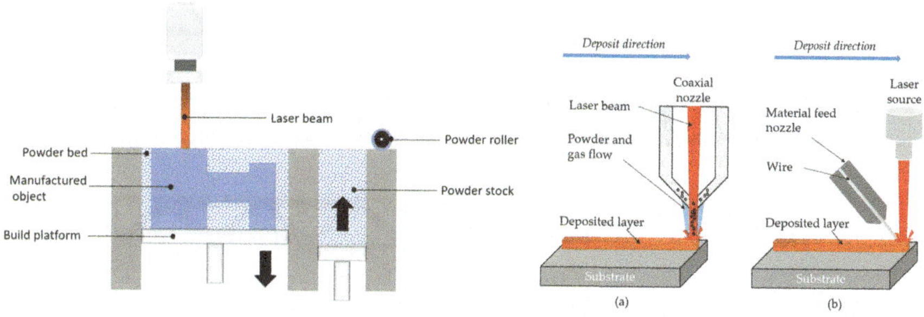

Fig. 1 Schematics of AM processes of (**left**) the PBF process [25], and (**right**) the DED process using (**a**) powder and (**b**) wire feedstock [19]. Both schematics are licensed under CC BY 4.0

2.2 Structural Optimisation

The advantages of AM methods have led to the creation of more efficient structures for critical aerospace applications. As structures with complex geometry and internal features can now be manufactured, methods of structural optimisation have become widely utilised and implemented. Structural optimisation refers to a technique whereby the material distribution of a structure is spatially changed to make the structure more efficient for a given purpose. Optimisation aims to address one or more objective functions, such as maximising stiffness, with respect to a given design constraint, such as minimising weight, while based on a set of defined loading conditions [39]. Numerical approaches through the finite element method (FEM) are often used to perform these tasks. For this, computational models are created and discretised to produce a set of connected nodes and elements with identical loading and boundary conditions as experienced in its application. By replicating the structural features and prescribed loading conditions in a computational domain, the load transfer pathways can be identified and regions of minimal load transfer can be removed to result in a more efficient structure.

Structural optimisation can be categorised into three primary areas; size, shape and topology, as displayed in Fig. 2, with associated levels of complexity in each [36]. Size optimisation is the most simple method that modifies the dimensions of existing parameters, such as length or thickness, to determine the optimal size of the current structure dimensions. Shape optimisation is more complex and works by shifting current nodal positions to modify the existing shape [36]. Topology optimisation is a more advanced method of optimisation that spatially optimises the distribution of material within a defined domain while considering design objectives [10]. In this method, elements and nodes can be removed whilst maintaining the load-bearing path elements, resulting in a unique, optimal solution [8]. Several objectives can also be applied simultaneously which results in a multiple-objective optimisation study, as well as combinations of various optimisation methods to refine the final design, such as performing an initial topology optimisation and refining the final design using shape optimisation. More unique approaches such as

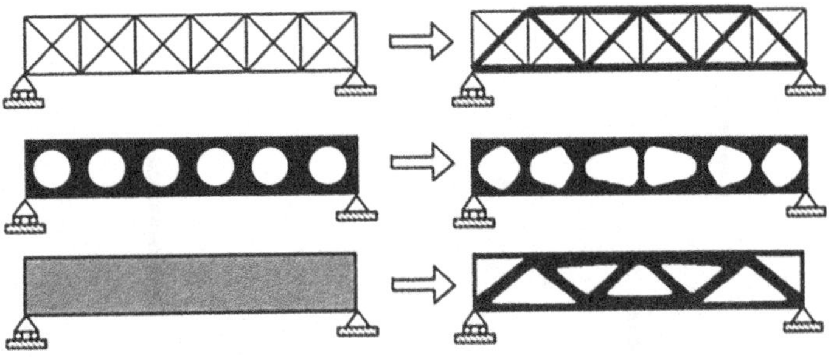

Fig. 2 Examples of the categories of structural optimisation including (top) size optimisation (middle) shape optimisation and (bottom) topology optimisation, presenting the (left) initial design and (right) optimised structure [38]. Licensed under CC BY 4.0

cellular or lattice structures can also be used to take full advantage of AM abilities, where individual designs can be created based on their structural functionality [26]. Various types of lattice structures can be used with numerous properties, such as size and density, that can be tuned to their particular application.

3 Advantages of Additive Manufacturing Towards Efficient Structures

The implementation of design optimisation with the realised potential of AM technology has become increasingly valuable to the aerospace industry. The synergy between structural optimisation and AM processes provides an ideal approach for the fabrication of innovative components that take full advantage of AM's unique potential [21]. With the numerous benefits provided by AM that push the boundaries of lightweight, efficient components, it is considered to be an exceptional candidate to progress the aerospace sector towards a more sustainable outlook [31]. In particular, the space industry is poised to gain the maximum benefit from this technology, where most structures are highly customised with low-volume part production [33]. The removal of limitations imposed by traditional manufacturing methods and the unparalleled design freedom achievable makes AM an integral fabrication technique for the production of future structures. AM technology demonstrates several advantages towards the creation of more efficient structures that are outlined in this section, including weight-conscious designs, reduced material processing and energy requirements, decentralised manufacturing abilities and production flexibility.

3.1 Weight Reduction

The weight of spacecraft structures is a critical quality for mission operations. One of the main advantages that arise from structural optimisation is the ability to fabricate components with less material, thus reducing the weight of the structure, whilst maintaining critical strength characteristics. Aerospace applications are highly sensitive to mass, being directly proportional to the associated fuel requirements for the intended mission. Large structures, such as aircraft and rockets, consume ample fuel during their operations to transport their payload safely. In the case of orbital launches, multiple stages are often used with separate fuel sources, being a primary contributor to the total mass of the system [7]. This ensures enough thrust is provided for the vehicle to reach orbit safely whilst overcoming related drag and gravitational resistances with increasing altitude and velocity. The propulsion of rockets is achieved using a fuel source with an oxidiser, collectively making a propellant [32]. Both solid and liquid propellants are used in rockets, each producing a variety of harmful emissions during combustion [7]. These emissions can include water vapour (H_2O), nitrogen oxides and other pollutants such as alumina particles and carbon dioxide (CO_2) [32]. Each of these constituents can cause significant harm to the surrounding environments and atmosphere, with the accumulative effects being evident through the depletion of the ozone layer [18].

A paradox that arises from orbital launches is the relation of the fuel required to propel the vehicle. To launch the mass of the vehicle, including the fuel, a finite amount of fuel is required. As the mass of the vehicle system increases, additional fuel is required. This results in an increased supply of fuel to support the current mass at launch. Not only does that imply a larger environmental burden with more fuel combustion but also alludes to the benefits of a reduced launch weight. The Tsiolkovsky rocket equation, as denoted in Eq. 1, represents the numerical interpretation of an ideal change in rocket exhaust velocity to its mass ratio, assuming no atmospheric drag and change in thrust [33]. The change in the velocity (Δv) can be defined as the product of the exhaust velocity (v_e) and the natural logarithm of the ratio between the initial system mass including fuel (m_0) and final mass without propellant (m_f).

$$\Delta v = v_e \ln\left(\frac{m_0}{m_f}\right) \tag{1}$$

Through a graphical representation of this equation, presented in Fig. 3, it is evident that an increasing mass ratio $\left(\frac{m_0}{m_f}\right)$ increases the exhaust velocity ratio $\left(\frac{\Delta v}{v_e}\right)$. Assuming constant fuel usage and exhaust velocity, it can be derived from Eq. 1 that a lower initial mass (higher mass ratio) can result in a larger change in velocity. This signifies the inherent benefits of lightweight components for spacecraft vehicles.

The aforementioned benefits of AM with structural optimisation can provide significant advantages towards the creation of lightweight structures. Through optimisation techniques, structures can be redesigned in a manner that can significantly reduce their weight compared to their conventional counterpart whilst still

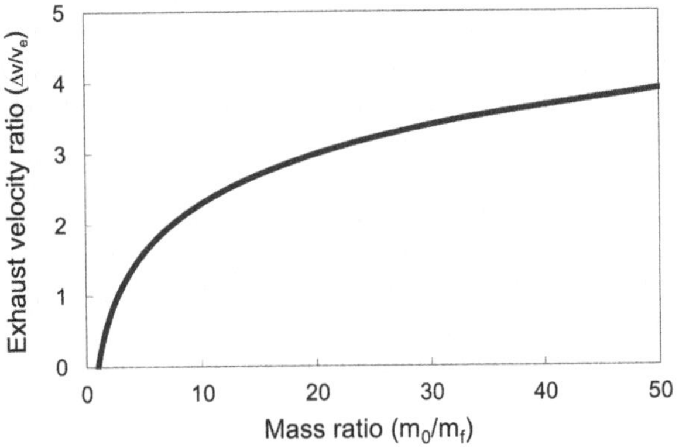

Fig. 3 Graphical representation of the Tsiolkovsky rocket equation

maintaining critical mechanical properties [15]. This has already been demonstrated in a variety of aerospace applications including satellite antennas, brackets and CubeSat designs [4]. Figure 4 depicts a case study performed for the optimisation of a star tracker camera bracket through generative design, where a weight saving of 79% was achieved that reduced the component mass from an initial 425 g to 89 g [24]. The achievable weight savings through optimal designs can lead to a reduction in the associated fuel requirements on missions, which in turn reduces the associated environmental impact.

Aside from sustainability benefits, economic advantages can also transpire from lightweight designs. While structural mass, fuel savings and emissions are the main focal points of optimal design considerations, the allocation of weight in orbital vehicles must also be considered. Opposed to reducing both the mass and fuel of the vehicle, the mass savings achieved by optimal structures can be translated into excess fuel supplied for the mission. Whilst this ensures the mass of the system does not change, the additional fuel can be used to increase mission duration or distance, which is also important for space operations. Conversely, the weight savings achieved could be converted into additional payloads on spacecraft, meaning that more data can be collected or experiments performed in the same mission profile. Both of these abilities can be economically advantageous. With the cost of space launches known to exceed €10,000 per kilogram, maximising the capacity of their current vehicle provides a more efficient solution as opposed to multiple orbital launches [20].

3.2 *Reduced Material Processing and Energy Requirements*

The effective allocation of material in components through optimised designs can further reduce the associated material processing and energy requirements through AM methods. Traditional, subtractive manufacturing techniques typically remove material from a bulk material source via processes such as milling or turning. This not only generates significant material wastage but the processing itself can also

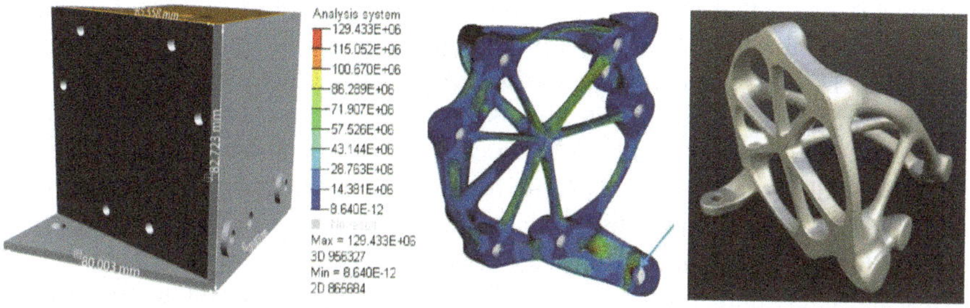

Fig. 4 Optimisation process of a star tracker camera bracket including (left) the original structure, (centre) the FEM stress distribution of the optimised design, and (right) the optimised component fabricated through AM [24]. Licensed under CC BY 4.0

include high energy requirements and costs through large machinery usage. Contrastingly, AM fabricates structures through a sequential layering process with high material efficiency. This allows structures to be realised without the aid of moulds or other casting additions, along with minimal material wastage. High buy-to-fly ratios can also be achieved, which relates the initial material required to the material used in the application. In subtractive processes, this can exceed almost 20:1, compared to the achievable 1:1 for AM techniques [15].

In addition to the high material fabrication efficiency, AM processes have the added benefit of high material recyclability. Whilst the material removed in subtractive processes can often be recycled through remelting, such as for metals, other material categories are much more difficult to recycle. This is particularly the case for thermoset composites, such as carbon fibre composites that use an epoxy matrix system [40]. In powder-based AM processes, such as powder-bed fusion, the unmelted feedstock can regularly be recycled and reused [34]. After a fabricated part is removed from the printing bed, the excess powder can be put through a filtration system to remove unusable or partially melted, large particles. The remaining powder can produce a quantity of feedstock that can be readily added back to the bulk feedstock supply and used for the next AM print. Such filtration processes can recycle 95–98% of unmelted feedstock, which is unparalleled in other manufacturing industries [27].

3.3 Decentralised Manufacturing

The nature of AM technology provides the notion of decentralised manufacturing to advance aerospace structures towards a more sustainable outlook. AM enables the ability to reduce the associated supply chain process through in-house manufacturing facilities and part consolidation. The construction of large production jobs is often tasked to a manufacturer, generally more than one, which has the required labour or machinery capabilities. The assembled structure is then transported to the designated buyer, either internationally or domestically. Not only can this result in potential project delays, costs and setbacks, but also increases the associated environmental impact that arises from the transportation. AM processes can reconfigure this supply chain by fabricating parts locally at the final destination using a single supplier [11]. With an increase in access to AM facilities across commercial industries and institutions, the need for extensive supply chains can be reduced and can create more economical solutions to manufacturing difficulties.

Additional advantages can also arise from this, including more time-efficient and cost-effective manufacturing through an on-demand approach. During project timelines, substantial planning is required to ensure components are fabricated on time to test and safely implement as part of a larger assembly. As a result, many items are often purchased in advance to avoid later delays, thus creating a populous inventory of large structures and components. Using AM, components can be fabricated only when needed, reducing the requirement to maintain bulk storage and the risk of wastage as older component models phase out [2]. The notion of on-demand manufacturing also provides several benefits to service and maintenance. A damaged component traditionally requires either repair or replacement, depending on the extent. Obtaining replacements for uncommon components is often costly and timely, creating significant inefficiency for a low number of parts. AM technology enables replacement components to be manufactured on-demand using localised machines where the quantity is less of a concern. As an alternative to replacement, AM technology can also be used to locally repair structures through methods such as cold spray or cladding techniques.

3.4 Production Flexibility

AM technology enables significant production flexibility through prototyping ability and ease of component modification. Additive processes initially construct a 3-dimensional model through computer-aided design (CAD) software and then manufacture them to a near net or final shape [4]. This is a complex process that requires technical ability and an understanding of various design processes, particularly that of the design for additive manufacturing (DFAM) paradigm. Due to the sequential layering process of AM, knowledge of the inherent manufacturing limitations is necessary to create functional components that can also be manufactured additively. This can include understanding the effects of manufacturing defects, printing parameters and build orientations [33]. DFAM logic aims to benefit the AM process by considering a range of manufacturing and design aspects to improve the functional performance of a part [12]. The mechanical properties of AM components have shown to be highly related to their microstructural properties and are a large concern in the design of components. Material anisotropy may be prevalent in AM components, where the mechanical properties of a component can vary depending on the orientation it was fabricated in, whether in the horizontal plane, perpendicular to the build direction, or in the vertical plane, parallel to the build direction [41]. Comprehending the effect of build orientations and overhang angles is an essential component of DFAM logic. With AM technology, prototypes can be created that use various DFAM considerations which can be analysed to determine the most optimal manufacturing parameters for that given component.

To assess the performance and quality of structures, scaled prototypes can also be manufactured and tested. Using this as a proof-of-concept, modifications can be made to the prototype designs and tested under comparative loading conditions. This is not only faster than manufacturing full-scale models but can be more cost-effective than an entire re-design after fabrication [11]. The direct link between CAD software and component manufacture further contributes to the production flexibility of AM technology. Dimensional tolerances of structures are critical to complex systems to ensure they are suitable as part of a larger assembly. Subtractive processes, particularly casting, often rely on a mould that can be used repeatedly to form the component geometry [9]. With the complex geometry often associated with AM, component dimensions can be modified with relative ease. These files can also be shared with associated parties on collaborative projects to be viewed and altered where necessary. Ensuring the proposed structure is acceptable to all parties in a computational domain is essential before full-scale manufacturing to avoid unnecessary rework and error.

4 Current Challenges and Limitations

4.1 Quality Assurance and Certification Standards

Despite the numerous advantages of AM technology, several challenges and limitations impede the widespread implementation of this technology in industry applications. For AM to be considered a safe and robust technology for future aerospace applications, it must evolve into a more versatile technology with increased reliability. Quality assurance and certification standards are currently some of the primary challenges in the development of AM, with the quality and repeatability of components being identified as some of the major flaws [35]. Particularly in aerospace applications where components can be load-bearing and critical to structural health, AM must be able to demonstrate suitable quality in fabricated components with repeatable results. The inherent defects produced during typical AM processes can significantly affect the properties of fabricated components, making their quality assurance difficult to measure. This is further heightened by each component possessing a unique set of defects and hence its own unique mechanical properties, where common standard and verification techniques used for subtractive and cast components cannot directly be applied in the same way to AM [33].

Porosity and void defects are inevitable in AM components, with their elimination the focus of much research that still presents several difficulties [9]. Defects can arise from various sources including gaseous pockets being created due to incomplete melting of feedstock and variations in powder feedstock size resulting in voids. These defects can initiate significant problems to the structural assurance of components by affecting their mechanical strength and fatigue life. Porosity defects have been known to facilitate crack initiation during mechanical loading by creating a unique stress state through the structure [5]. Surface roughness is also a concern in AM components, being one of the primary factors in the fatigue performance of structures. Surface roughness can be caused by manufacturing process parameters including the size of particle feedstock and printing parameters such as layer thickness [23]. Both porosity and surface roughness can act as sites of failure initiation which are difficult to not only predict but, in the case of internal porosity, challenging to observe [1].

4.2 Part Scalability and Production Rate

The limitations of manufacturing size with current AM technology also present several challenges to the widespread use of AM in the aerospace industry. Current AM build chambers have a finite size in their build platform and volume envelope, limiting parts of constrained size to be manufactured [37]. In addition, the scalability of parts can also impede the manufacturability of components. AM parts cannot always simply be scaled to be fabricated as they can exceed build size limits, where redesigning greater build platforms and capabilities is necessary to achieve larger scalability. However, the current issues with printing quality and associated defects will still be an inherent concern and can amplify these effects on larger components.

As each layer is fabricated individually in AM processes, the total fabrication time can be slower in overall build cycles compared to alternative manufacturing techniques. In accompaniment with their finite build size, large fabrication jobs that require several components could take longer to manufacture. AM excels in part customisation with smaller production rates, where fewer components are created but highly tuned and optimised to their intended application. This makes the space industry poised to gain maximum benefit from this technology. However, the limitations of smaller production rates can affect the widespread adoption of AM for common aerospace structures required in larger quantities, such as aerospace brackets. Developing larger build capabilities that can accommodate greater dimensions will also be essential towards the broader implementation of AM. Aerospace structures that will benefit from the customisation of larger parts can include wing structures with included ribs and stringers fabricated as one part or engine structures with integrated cooling channels [17].

5 Future Opportunities

As the aerospace industry continues to develop, the use of AM technology will be essential for future operations. With increased interest in space exploration and research, future missions will aim to explore deeper into space with significantly longer durations. However, developing innovative technologies that can reliably perform on these missions is crucial to achieving such feats. The unique characteristics of AM technology have already reimagined how components are designed and provide several benefits towards more efficient aerospace structures, with further opportunities prevalent by implementing AM in various life cycle stages. The unprecedented nature of AM technology has further led to its ingenious use for in-space manufacturing capabilities and ISRU. To add to the current advantages, both of these uses of AM can increase the efficiency of orbital launches and have the ability to enable longer mission durations.

5.1 Integration of AM in Various Life Cycle Stages

Improving the integration of AM technology with other manufacturing techniques and throughout various life cycle stages can enhance the implementation of AM for aerospace applications. AM technology has already demonstrated the benefits of advanced design methods such as DFAM and optimisation techniques in creating efficient aerospace structures. However, implementing AM throughout various life cycle stages can also be valuable towards increasing the sustainability of the aerospace sector. As presented in [11], Fig. 5 shows the potential opportunities to identify the sustainability benefits of AM across several life cycle stages. This encompasses several main areas of product development where AM technology has opportunities to be implemented in the aerospace sector. It is noted that these life cycle stages incorporate closed-loop processes, meaning that the implementation of AM technology in one area can provide benefits at other stages of component life, such as for the repair of structures to continue service. As AM fields continue to develop, such as through novel material, higher quality fabrications or innovative design techniques, the life cycle of components can also benefit. Creating a closer relationship within the AM process will further contribute to this, such as improving the component-technology relation between CAD, simulation and fabricated structures. Improving the ability to predict the performance of AM structures before fabrication can lead to greater reliability and the understanding of the limitations of fabricated components which can assist in the end-of-life procedure required to safely repair or replace components [22].

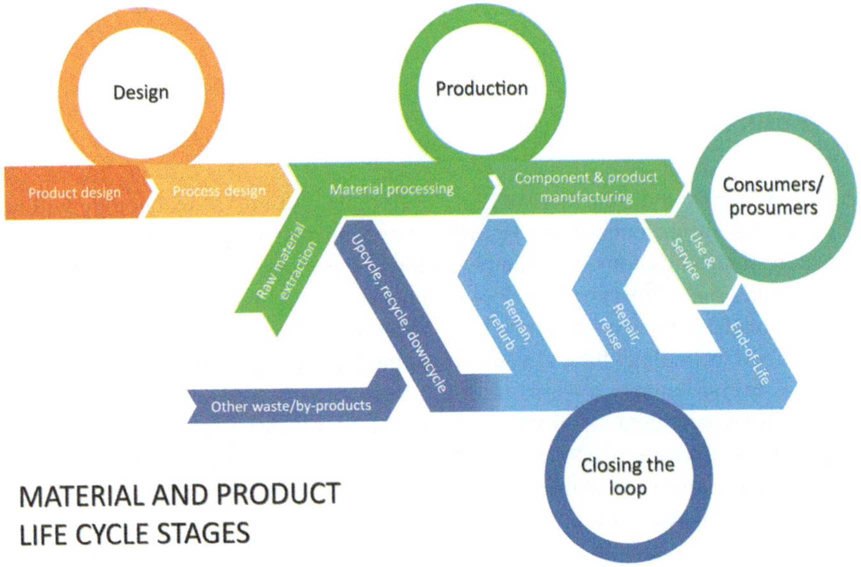

Fig. 5 Opportunities for AM material and product implementation across several life cycle stages presented by Ford and Despeisse [11]. Licensed under CC BY 4.0

5.2 *In-Space Manufacturing Capabilities*

With the inherent difficulties associated with space launches, particularly large costs, challenging size constraints and limited fuel supply, finding innovative solutions to extend mission durations is necessary. One method can include increasing the size and capacity of current vehicles to accommodate more fuel and larger payload structures, such as satellites. However, this is not a sustainable method to progress space operations as more materials and manufacturing will be required to integrate larger structures, along with increased fuel requirements to support the additional weight. As aforementioned, these factors already contribute significantly to the ecological impact of aerospace operations. In contrast, AM technology enables the unique ability to fabricate structures directly in space which presents an abundance of opportunities for novel spacecraft development and sustainment concepts.

Rather than developing new vehicles for larger payloads, material feedstock can be supplied for launch that can be used with onboard AM machines to fabricate components in situ. The notion of such technology provides several advantages for future mission capabilities. AM feedstock can maximise volume efficiency in current spacecraft vehicles by being compactly stored as opposed to larger structures that require greater dimensional space and volume. Parts such as tools and other items can be fabricated on-demand, without the need to resupply from other space platforms or Earth, to create a more self-sustaining system [20]. These components also have the added potential to be reused or recycled to produce new components using refabricator systems [33]. Once components are no longer needed, the material can be reused in AM processes to create other components, resulting in minimal waste and maximising the material's efficiency.

Aside from the economic advantages of high material and processing efficiency, the creation of novel structures post-launch is also beneficial. Space launches experience high acceleration and vibrational forces that can have adverse effects on structural components, particularly the exposed outer systems [3]. Validation tests are often performed to ensure structural systems will not fail under such dynamic launch conditions. Precaution is taken during the design of components by adding additional material support as protection during the launch phase to increase the safety factor of the design. This provides limitations on the designs achievable to account for high loads experienced at the cost of additional weight. With the aid of AM, sensitive and ultra-lightweight structures can be fabricated once in space, eliminating the requirement to survive the harsh conditions of launch by remaining in a compact feedstock form [33]. Intricate designs, such as highly optimised and lattice structures, can be created that are suited only for the conditions of space and orbit, which are significantly less rigorous.

Whilst in-space manufacturing capabilities have already been demonstrated for small parts, such as the AM facility onboard the International Space Station (ISS), the fabrication of larger, more primary structures is yet to be explored in depth [42]. Space stations, much like the ISS, have the unique ability to represent not only a destination for in-space manufacture by producing parts onboard but can also act as a development platform for novel AM cabilities. CubeSat structures have the potential to be entirely manufactured on space stations and then launched, being able to explore further into space than current missions. Multiple structures fabricated through a similar nature have the potential to form a constellation system entirely created through in-space manufacturing capabilities. With material efficiency paramount in space operations, the benefits of reducing mission costs and exploring innovative design concepts can embark on a new generation of spacecraft structures that are unable to be created on Earth to survive launch into space.

5.3 *In Situ Resource Utilisation*

Stemming from the notion of in-space manufacturing, the ability of ISRU with AM technology further presents several opportunities for more efficient space operations and structures. As operations begin to extend towards longer-duration interplanetary missions, the creation of large infrastructure for habitation purposes will be required. The concept of ISRU aims to utilise available resources in local environments as a method for self-sustaining construction [14]. Natural regolith already found in abundance on inter-terrestrial planets can be used as the feedstock supply for AM processes to fabricate structures. The main advantages of this are overcoming the costly launch constraints and available resources for extended missions. Rather than transporting material feedstock for habitation construction, they can be locally sourced at their destination to reduce not only the overall risk but also the associated cost of launch [16]. Therefore, only the AM fabrication system is required to be included for space missions and deployed at the destination. By eliminating the feedstock at launch, the weight of launch vehicles can be further reduced, leading to more efficient launches with an overall reduced environmental impact.

6 Conclusions

The unique characteristics of AM technology demonstrate great potential for significant advancements towards a more efficient and sustainable aerospace sector. The nature of AM processes and their close link to integrate structural optimisation techniques presents several advantages for the creation of innovative structural designs. The capability for weight reduction is one of the primary benefits provided by AM technology, not only to decrease the associated weight of aerospace structures but to further minimise the material processing and energy requirements. The ability of localised manufacturing with a reduced supply chain also allows for a more cost and time-effective manufacture, along with the production flexibility enabled through prototyping ability. Despite the numerous advantages, there are still several challenges associated with AM that inhibit its widespread adoption.

Particularly with quality assurance creating components with repeatable results that can suffice certification standards. Additionally, the current limitations of build size achievable with AM techniques and production rate presents areas to be addressed for larger aerospace structures to be fabricated. Stemming from the advantages, resourceful methods of in-space manufacturing and ISRU can provide an abundance of future opportunities to explore the full potential of AM. The prospects to integrate AM technology throughout various life cycle stages further allow AM to contribute towards the development and maintenance of aerospace structures. As research into the aerospace and space domains continues to evolve, the use of AM technology with inevitably presents itself as an effective pathway to progress the sector towards a more sustainable outlook.

References

1. Afroz, L., Inverarity, S. B., Qian, M., Easton, M., & Das, R. (2023). Analysing the effect of defects on stress concentration and fatigue life of L-PBF AlSi10Mg alloy using finite element modelling. *Progress in Additive Manufacturing*. https://doi.org/10.1007/s40964-023-00457-0
2. Attaran, M. (2017). The rise of 3-D printing: The advantages of additive manufacturing over traditional manufacturing. *Business Horizons, 60*(5), 677–688. https://doi.org/10.1016/j.bushor.2017.05.011
3. Blachowicz, T., Ehrmann, G., & Ehrmann, A. (2021). Metal additive manufacturing for satellites and rockets. *Applied Sciences, 11*(24). https://doi.org/10.3390/app112412036
4. Blakey-Milner, B., Gradl, P., Snedden, G., Brooks, M., Pitot, J., Lopez, E., Leary, M., Berto, F., & du Plessis, A. (2021). Metal additive manufacturing in aerospace: A review. *Materials & Design, 209*, 110008. https://doi.org/10.1016/j.matdes.2021.110008
5. Borgue, O., Panarotto, M., & Ola, I. (2018). *Impact on design when introducing additive manufacturing in space applications*. Paper presented at the international design conference – Design 2018,
6. D'Angelo, O., Kuthe, F., Liu, S.-J., Wiedey, R., Bennett, J. M., Meisnar, M., Barnes, A., Kranz, W. T., Voigtmann, T., & Meyer, A. (2021). A gravity-independent powder-based additive manufacturing process tailored for space applications. *Additive Manufacturing, 47*, 102349. https://doi.org/10.1016/j.addma.2021.102349
7. Dallas, J. A., Raval, S., Alvarez Gaitan, J. P., Saydam, S., & Dempster, A. G. (2020). The environmental impact of emissions from space launches: A comprehensive review. *Journal of Cleaner Production, 255*, 120209. https://doi.org/10.1016/j.jclepro.2020.120209
8. Das, R., & Jones, R. (2011). Topology optimisation of a bulkhead component used in aircrafts using an evolutionary algorithm. *Procedia Engineering, 10*, 2867–2872. https://doi.org/10.1016/j.proeng.2011.04.476
9. Dordlofva, C., & Törlind, P. (2017). Qualification challenges with additive manufacturing in space applications. In *2017 International Solid Freeform Fabrication Symposium, 2017*. University of Texas at Austin, pp 2699–2712.
10. Esfarjani, S. M., Dadashi, A., & Azadi, M. (2022). Topology optimization of additive-manufactured metamaterial structures: A review focused on multi-material types. *Forces in Mechanics, 7*, Art no 100100. https://doi.org/10.1016/j.finmec.2022.100100
11. Ford, S., & Despeisse, M. (2016). Additive manufacturing and sustainability: An exploratory study of the advantages and challenges. *Journal of Cleaner Production, 137*, 1573–1587. https://doi.org/10.1016/j.jclepro.2016.04.150

12. Haruna, A., & Jiang, P. (2022). Adaptability analysis of design for additive manufacturing by using fuzzy Bayesian network approach. *Advanced Engineering Informatics, 52*, 101613. https://doi.org/10.1016/j.aei.2022.101613
13. Huang, Y., Leu, M. C., Mazumder, J., & Donmez, A. (2015). Additive manufacturing: Current state, future potential, gaps and needs, and recommendations. *Journal of Manufacturing Science and Engineering, 137*(1), 014001. https://doi.org/10.1115/1.4028725
14. Isachenkov, M., Chugunov, S., Akhatov, I., & Shishkovsky, I. (2021). Regolith-based additive manufacturing for sustainable development of lunar infrastructure—An overview. *Acta Astronautica, 180*, 650–678. https://doi.org/10.1016/j.actaastro.2021.01.005
15. Joshi, S. C., & Sheikh, A. A. (2015). 3D printing in aerospace and its long-term sustainability. *Virtual and Physical Prototyping, 10*(4), 175–185. https://doi.org/10.1080/17452759.2015.1111519
16. Karl, D., Cannon, K. M., & Gurlo, A. (2022). Review of space resources processing for Mars missions: Martian simulants, regolith bonding concepts and additive manufacturing. *Open Ceramics, 9*, 100216. https://doi.org/10.1016/j.oceram.2021.100216
17. Kerstens, F., Cervone, A., & Gradl, P. (2021). End to end process evaluation for additively manufactured liquid rocket engine thrust chambers. *Acta Astronautica, 182*, 454–465. https://doi.org/10.1016/j.actaastro.2021.02.034
18. Kokkinakis, I. W., & Drikakis, D. (2022). Atmospheric pollution from rockets. *Physics of Fluids, 34*(5), 056107. https://doi.org/10.1063/5.0090017
19. Lee, J., Park, H. J., Chai, S., Kim, G. R., Yong, H., Bae, S. J., & Kwon, D. (2021). Review on quality control methods in metal additive manufacturing. *Applied Sciences, 11*(4), 1966. https://doi.org/10.3390/app11041966
20. Mitchell, A., Lafont, U., Hołyńska, M., & Semprimoschnig, C. (2018). Additive manufacturing — A review of 4D printing and future applications. *Additive Manufacturing, 24*, 606–626. https://doi.org/10.1016/j.addma.2018.10.038
21. Mohd Yusuf, S., Cutler, S., & Gao, N. (2019). Review: The impact of metal additive manufacturing on the aerospace industry. *Metals, 9*(12), 1286. https://doi.org/10.3390/met9121286
22. Mostafaei, A., Zhao, C., He, Y., Reza Ghiaasiaan, S., Shi, B., Shao, S., Shamsaei, N., Wu, Z., Kouraytem, N., Sun, T., Pauza, J., Gordon, J. V., Webler, B., Parab, N. D., Asherloo, M., Guo, Q., Chen, L., & Rollett, A. D. (2022). Defects and anomalies in powder bed fusion metal additive manufacturing. *Current Opinion in Solid State and Materials Science, 26*(2), 100974. https://doi.org/10.1016/j.cossms.2021.100974
23. Nakatani, M., Masuo, H., Tanaka, Y., & Murakami, Y. (2019). Effect of surface roughness on fatigue strength of Ti-6Al-4V alloy manufactured by additive manufacturing. *Procedia Structural Integrity, 19*, 294–301. https://doi.org/10.1016/j.prostr.2019.12.032
24. Orme, M., Madera, I., Gschweitl, M., & Ferrari, M. (2018). Topology optimization for additive manufacturing as an enabler for light weight flight hardware. *Designs, 2*(4), 51. https://doi.org/10.3390/designs2040051
25. Pal, R., & Basak, A. (2022). Linking powder properties, printing parameters, post-processing methods, and fatigue properties in additive manufacturing of AlSi10Mg. *Alloys, 1*(2), 149–179. https://doi.org/10.3390/alloys1020010
26. Panesar, A., Abdi, M., Hickman, D., & Ashcroft, I. (2018). Strategies for functionally graded lattice structures derived using topology optimisation for additive manufacturing. *Additive Manufacturing, 19*, 81–94. https://doi.org/10.1016/j.addma.2017.11.008
27. Petrovic, V., Vicente Haro Gonzalez, J., Jordá Ferrando, O., Delgado Gordillo, J., Ramón Blasco Puchades, J., & Portolés Griñan, L. (2010). Additive layered manufacturing: Sectors of industrial application shown through case studies. *International Journal of Production Research, 49*(4), 1061–1079. https://doi.org/10.1080/00207540903479786
28. Psanis, C., Triantafyllou, E., Giamarelou, M., Manousakas, M., Eleftheriadis, K., & Biskos, G. (2017). Particulate matter pollution from aviation-related activity at a small airport of the Aegean Sea Insular Region. *Science of the Total Environment, 596–597*, 187–193. https://doi.org/10.1016/j.scitotenv.2017.04.078

29. Qiu, R., Hou, S., Chen, X., & Meng, Z. (2021). Green aviation industry sustainable development towards an integrated support system. *Business Strategy and the Environment, 30*(5), 2441–2452. https://doi.org/10.1002/bse.2756
30. Rodrigues, J., Barter, S., & Das, R. (2024). The role of additive manufacturing towards sustainable aerospace structures. In T. H. Karakoc, R. Das, I. Ekmekci, A. Dalkiran, & A. H. Ercan (Eds.), *Green approaches in sustainable aviation* (pp. 107–114). Springer.
31. Rodrigues, J., Foster, W., Barter, S., & Das, R. (2023). *A comparison of metal additive structures to polymer composite structures for space applications*. Paper presented at the AIAC 2023: 20th Australian international aerospace congress, Melbourne
32. Ryan, R. G., Marais, E. A., Balhatchet, C. J., & Eastham, S. D. (2022). Impact of rocket launch and space debris air pollutant emissions on stratospheric ozone and global climate. *Earth's Futures, 10*(6), e2021EF002612. https://doi.org/10.1029/2021ef002612
33. Sacco, E., & Moon, S. K. (2019). Additive manufacturing for space: Status and promises. *The International Journal of Advanced Manufacturing Technology, 105*(10), 4123–4146. https://doi.org/10.1007/s00170-019-03786-z
34. Shanbhag, G., & Vlasea, M. (2021). Powder reuse cycles in electron beam powder bed fusion-variation of powder characteristics. *Materials (Basel), 14*(16), 4602. https://doi.org/10.3390/ma14164602
35. Tapia, G., & Elwany, A. (2014). A review on process monitoring and control in metal-based additive manufacturing. *Journal of Manufacturing Science and Engineering, 136*(6), 060801. https://doi.org/10.1115/1.4028540
36. Tejani, G. G., Savsani, V. J., Patel, V. K., & Savsani, P. V. (2018). Size, shape, and topology optimization of planar and space trusses using mutation-based improved metaheuristics. *Journal of Computational Design and Engineering, 5*(2), 198–214. https://doi.org/10.1016/j.jcde.2017.10.001
37. Thomas-Seale, L. E. J., Kirkman-Brown, J. C., Attallah, M. M., Espino, D. M., & Shepherd, D. E. T. (2018). The barriers to the progression of additive manufacture: Perspectives from UK industry. *International Journal of Production Economics, 198*, 104–118. https://doi.org/10.1016/j.ijpe.2018.02.003
38. Tsavdaridis, K. D., Kingman, J. J., & Toropov, V. V. (2015). Application of structural topology optimisation to perforated steel beams. *Computers & Structures, 158*, 108–123. https://doi.org/10.1016/j.compstruc.2015.05.004
39. Tyflopoulos, E., & Steinert, M. (2022). A comparative study of the application of different commercial software for topology optimization. *Applied Sciences, 12*(2), 611. https://doi.org/10.3390/app12020611
40. Yang, Y., Boom, R., Irion, B., van Heerden, D.-J., Kuiper, P., & de Wit, H. (2012). Recycling of composite materials. *Chemical Engineering and Processing: Process Intensification, 51*, 53–68. https://doi.org/10.1016/j.cep.2011.09.007
41. Zhu, J., Zhou, H., Wang, C., Zhou, L., Yuan, S., & Zhang, W. (2021). A review of topology optimization for additive manufacturing: Status and challenges. *Chinese Journal of Aeronautics, 34*(1), 91–110. https://doi.org/10.1016/j.cja.2020.09.020
42. Zocca, A., Wilbig, J., Waske, A., Günster, J., Widjaja, M. P., Neumann, C., Clozel, M., Meyer, A., Ding, J., Zhou, Z., & Tian, X. (2022). Challenges in the technology development for additive manufacturing in space. *Chinese Journal of Mechanical Engineering: Additive Manufacturing Frontiers, 1*(1), 100018. https://doi.org/10.1016/j.cjmeam.2022.100018

Artificial Intelligence and Machine Learning in Manufacturing of Aircrafts

Ryoichi Tsuzuki

1 Introduction

In the Fourth Industrial Revolution, "Digital Transformation (DX)" is being implemented. This transformation involves the digitization of all industrial information. In the aircraft industry, automated robots and IoT systems for data collection are being introduced to improve productivity. In addition, artificial intelligence technologies are being widely studied as shown in Fig. 1 [1]. Productivity improvement here is referred to as Productivity, Quality, Cost, Delivery, Safety, Morale, and Environment "PQCDSME."

Research on welding robots and inspection systems has focused on the incorporation of several sensors, including electrical current, voltage, and sensors to optimize welding and inspection conditions. The optimization system of production layout and schedule for manufacturing-related process have been also investigated and incorporated. Some optimization technologies which are categorized artificial intelligence technology has been applied to this system.

In this situation, the advanced manufacturing system was established [2]. At the aircraft industry, information network technology based on advanced manufacturing system concept was established. Firstly, an outline of advanced manufacturing system for innovative manufacturing activities is introduced, and then artificial intelligence for optimization is described. The digital smart factory constructionis summarized at the end.

R. Tsuzuki (✉)
Kawasaki Heavy Industries Ltd, Aerospace System Company, Tokyo, Japan

T. H. Karakoc et al. (eds.), *Sustainable Materials and Manufacturing Techniques in Aviation*, Sustainable Aviation, https://doi.org/10.1007/978-3-031-62987-7_4

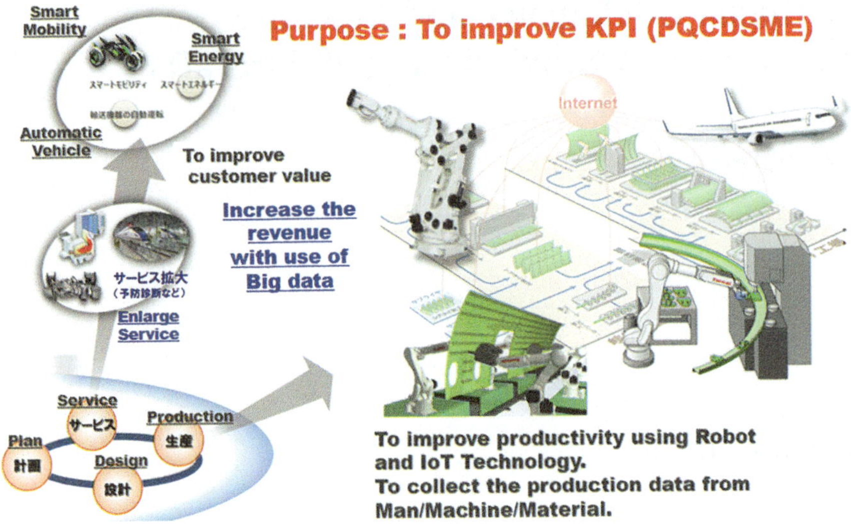

Fig. 1 Digitalization plan to improve productivity in the aircraft industry

2 Advanced Manufacturing System Overview

Advanced manufacturing system is established "Kaizen (improvement)" in below five steps as shown in Fig. 2.

Step 1 Creating the standard data by evaluating human activities and by processing the conditions status of amounts and products
Step 2 Visualizing the processing time and production status using graphs and images
Step 3 Extracting the subjects and problems from the human actions data, processing data, and equipment data
Step 4 Solving the subjects and problems by optimizing the worker assignment and process conditions.
Step 5 Analyzing the data after solving the subjects and problems

Advanced manufacturing system is developed by using artificial intelligence. Artificial intelligence is applied to "PLM (Product Lifecycle Management Line)," "SCM (Supply Chain Management Line)," "Cyber Physical System (CPS)," and "Digital Twin System (DPS)" (Fig. 3). In order to deploy and apply artificial intelligence to PLM and SCM along with above five steps as shown in Fig. 2.

Step 1 The status of operation people, equipment, and product must be quantified for data standardizing.
Step 2 Data from actual production processes are quantified and graphed for comparison with standardized data to achieve higher accuracy.

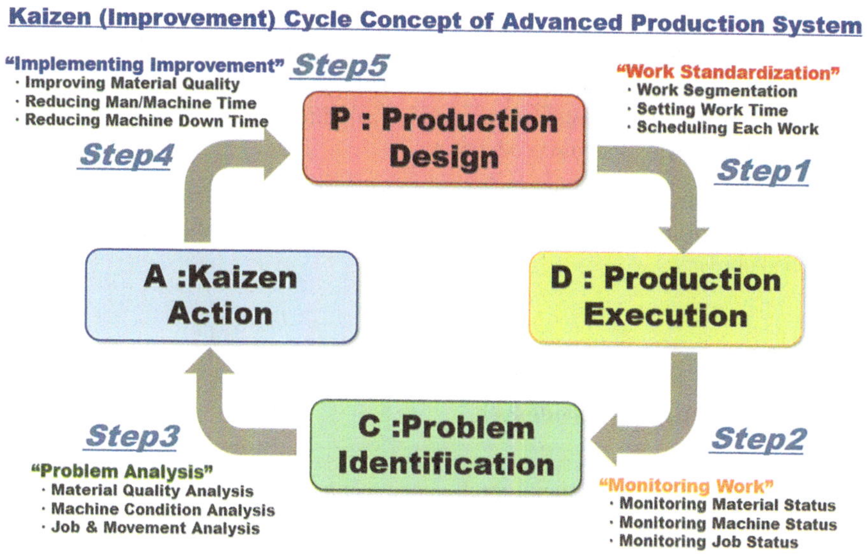

Fig. 2 Advanced production system as PDCA Kaizen (improvement) cycle

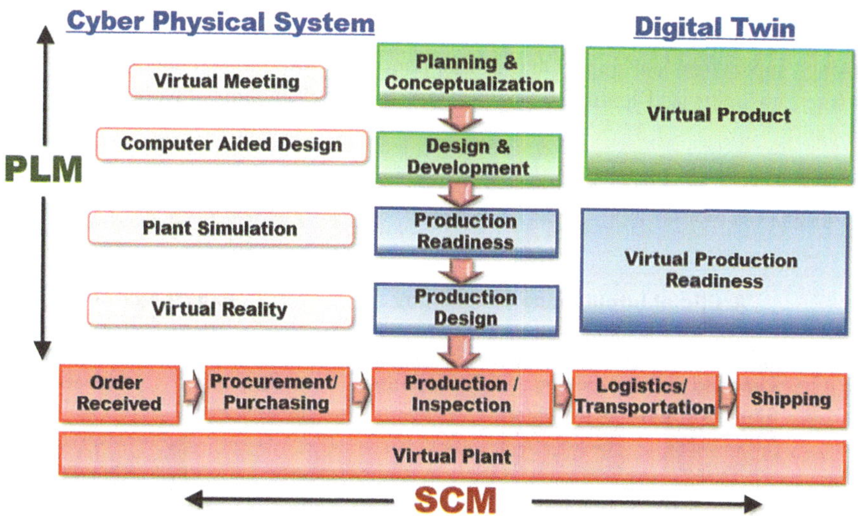

Fig. 3 Cyber physical system and digital twin as digitalized system for the line of PLM (product life cycle management) and SCM (supply chain management)

Step 3 From the comparative data in Step 2, express the phenomenon in a modeled equation. By inputting future boundary conditions from the modeled equation, problems in future manufacturing conditions are accurately extracted.

Step 4 To solve the identified problems, new technologies will be developed, and optimal manufacturing conditions and production plans will be established. For this, optimization operations using artificial intelligence will be applied.
Step 5 Apply new technologies, production conditions, and plans developed by artificial intelligence to actual production.

Subsequent sections will describe examples of advanced manufacturing systems in actual aircraft manufacturing, implemented according to the five-step approach described above.

3 Artificial Intelligence Map for Advanced Manufacturing Processes

The application map of artificial intelligence technology examples for the aircraft parts is shown in Fig. 4 [3]. Optimization and machine learning are considered the latest artificial intelligence technologies, which can be applied for actual manufacturing process through the advanced manufacturing systems.

Machine learning region consists of "statistical machine learning" and "deep learning." Both methods can be divided into "supervised learning" and "unsupervised learning." The output mapping functions are generated by the training data using "supervised learning." On the other hand, a learning model can be established using "unsupervised learning." "Supervised learning" includes "work analysis for

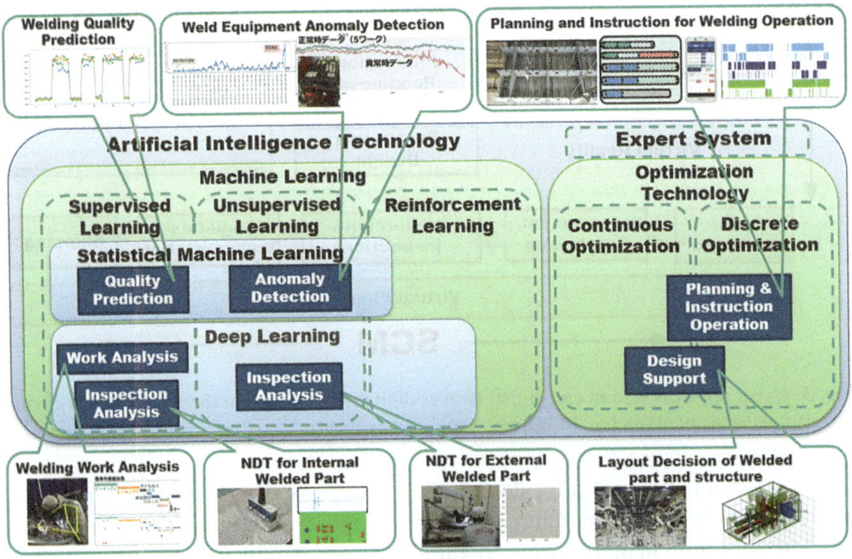

Fig. 4 Artificial intelligence technology map for welding and inspection process of aircraft manufacturing

welding operators" and "non-destructive inspection analysis for the inside of welds." "Unsupervised learning" includes "the non-destructive inspection like visual inspection of welds."

"Reinforcement learning" maximizes rewards by repeatedly searching for solutions. "Continuous optimization" can be used for "layout decision of welded parts." "Discrete optimization" is mainly applied to "planning and instruction for welding operation."

4 Development of Artificial Intelligence for Welding Operation

In the aero engine production, the optimum welding conditions have to been established. For these optimizations, the inspection testing results have to be evaluated. In the process of establishing welding conditions, parts, operator, and welding equipment conditions have to be digitized.

Figure 5 shows the digitalization results of TIG welded position strain. The strain is a factor in the occurrence of nonconformity. As a result, the strain could be evaluated visually between skilled operator and average operator by using three-dimensional scanner.

Next, the metallurgical structure was investigated for the cross section of the circumferential weld as shown in Fig. 6. In this investigation, several welding parameters were varied, such as weld heat input, welding speed, and welding time. Metallographic features and weld bead widths acquired by observation were converted into quantitative data by digitization.

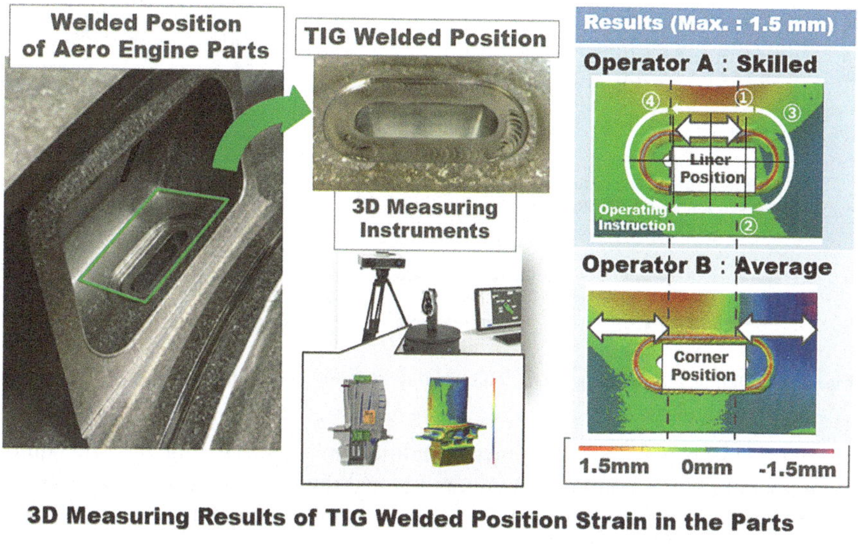

Fig. 5 Schematic illustration of the digitalization for the strain of TIG welded parts

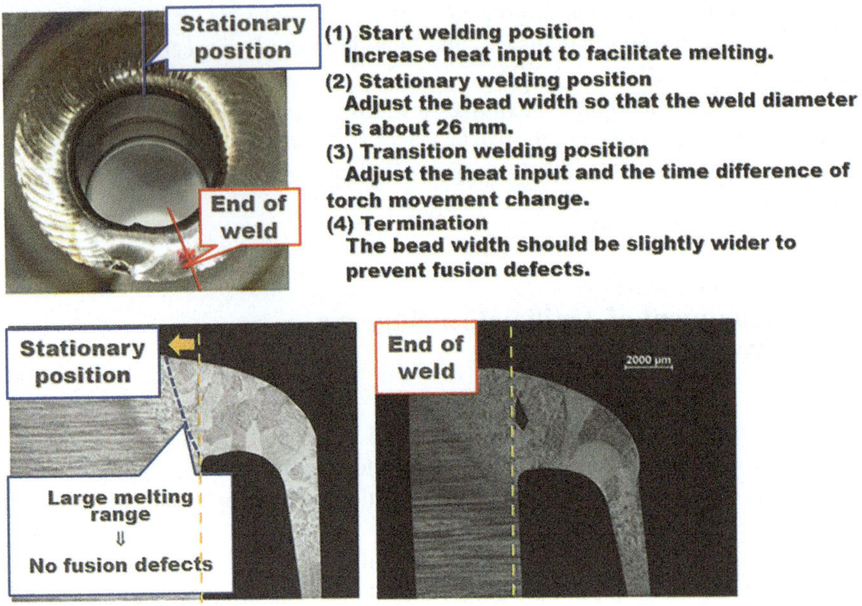

Fig. 6 Metallurgical evaluation results of TIG welded position

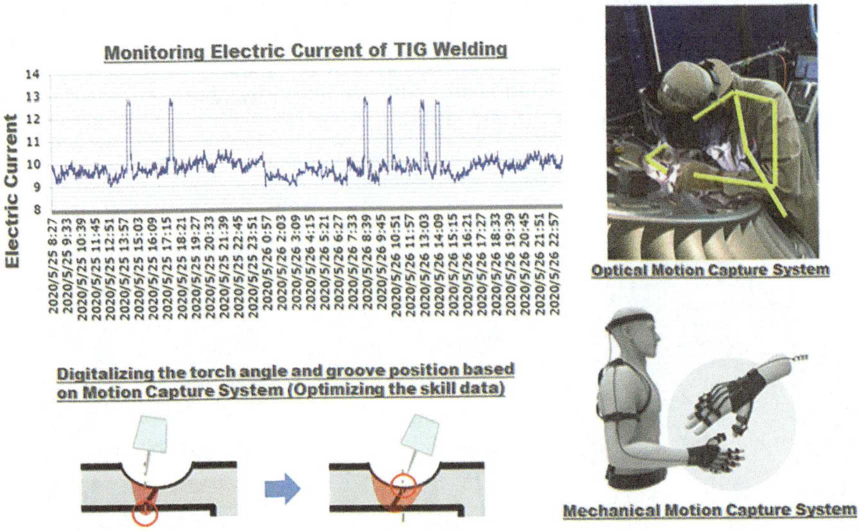

Fig. 7 The digitalization examples for TIG welding operator and equipment

TIG weld operator and equipment condition are digitized (Fig. 7). The equipment operating information are monitored in real time. The optical and mechanical motion capture technologies were developed and implemented for the standardization of motions of skilled operator.

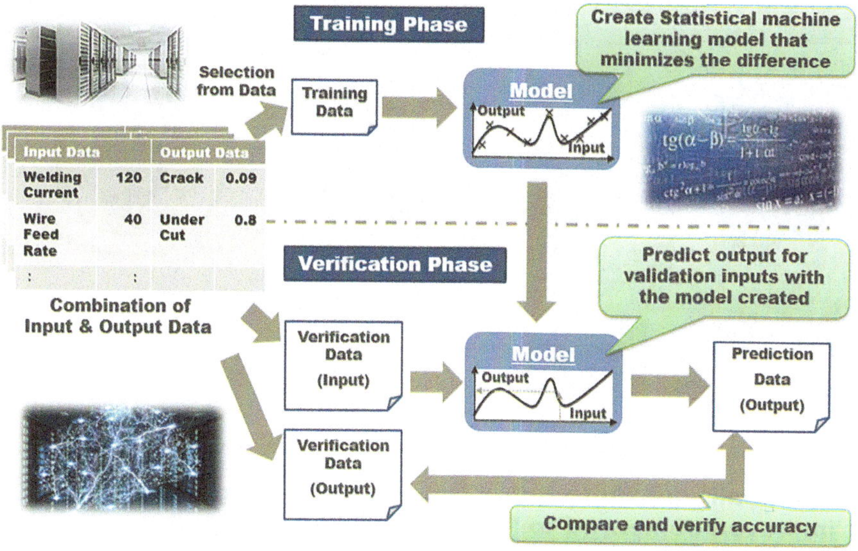

Fig. 8 Schematic illustration of statistical machine learning for TIG welding

Figure 8 shows the optimization method of standard conditions using the statistical machine learning. The welding operator condition and welding equipment conditions are obtained as digitized data. The statistical machine learning method is used to create the optimized model that minimizes the difference between input and output data in the training phase. The derived model is used to calculate and predict the output data from the input data in the verification phase.

Figure 9 shows the creation process of the optimized welding condition in TIG welding process. The calculation program is added to suppress the numerical deviation to the utmost limit in this TIG welding robot system. The robot can always perform welding under optimal conditions by assigning weight coefficients to important parameters and performing statistical machine learning is created.

As a result, the above automated TIG welding system can produce higher quality welds and the welding operation man-hours were reduced by 95%.

5 Development of Artificial Intelligence Technology for Judgement

X-ray system which was incorporated to automatic imaging and judgement system are described in the following. Figure 10 shows the robotized X-ray inspection system. In this system, the transmitted image is converted into a digital image. The flat panel can directly convert into electrical signals to display images. The use of film and photographing process was eliminated by developing this system.

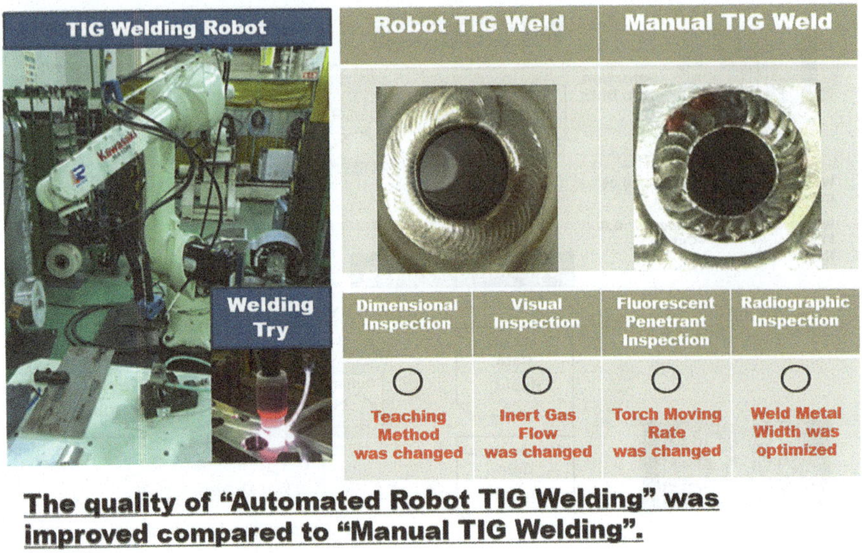

Fig. 9 Robot TIG welding system and inspection results of welded position

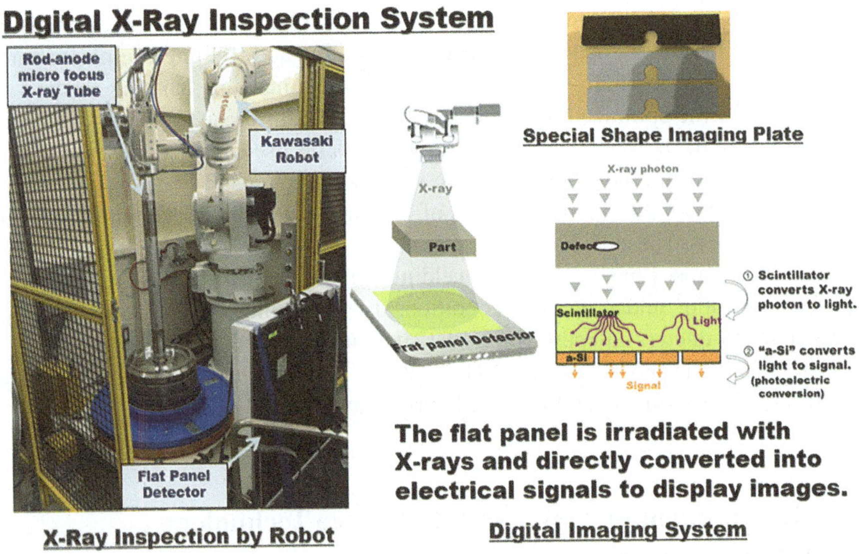

Fig. 10 Schematic illustration of digital X-ray inspection system

Figure 11 shows the inspection method for X-ray judgement using deep learning program. This deep learning method consists of "input layer," "multiple hidden layer," and "output layer." The machine learning method is used to create the neural network calculation model. This neural network model is used to judge the output data in verification phase.

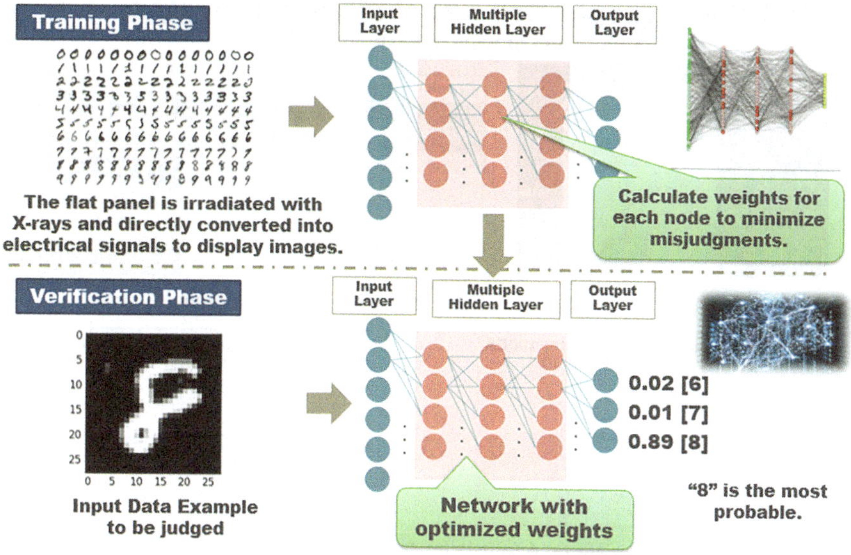

Fig. 11 Schematic illustration of deep learning for X-ray judgement

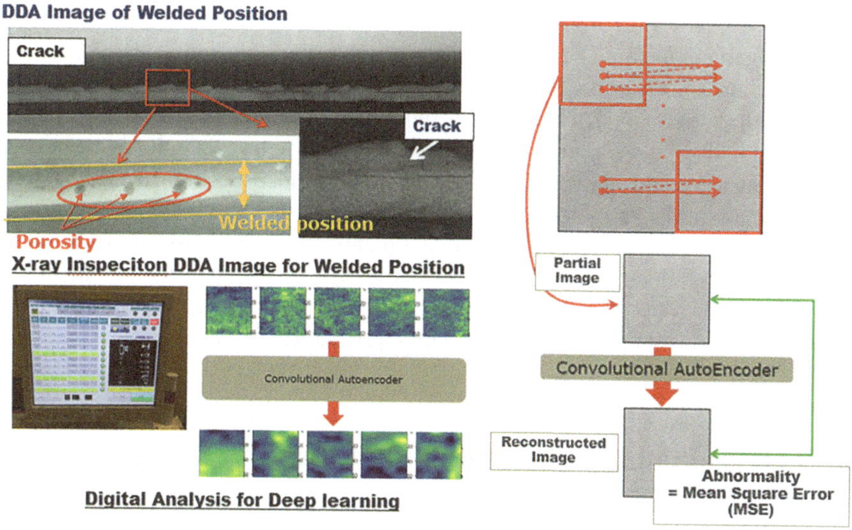

Fig. 12 Schematic illustration of digital X-ray judgement system

Figure 12 shows the automatic X-ray inspection judgement system using deep learning method. The digital partial images of welded position are acquired by using electrical signals. In Convolutional Auto-Encoder, the image is reconstructed as digital signal. The degree of abnormality "Mean Square Error (MSE)" in the reconstructed image.

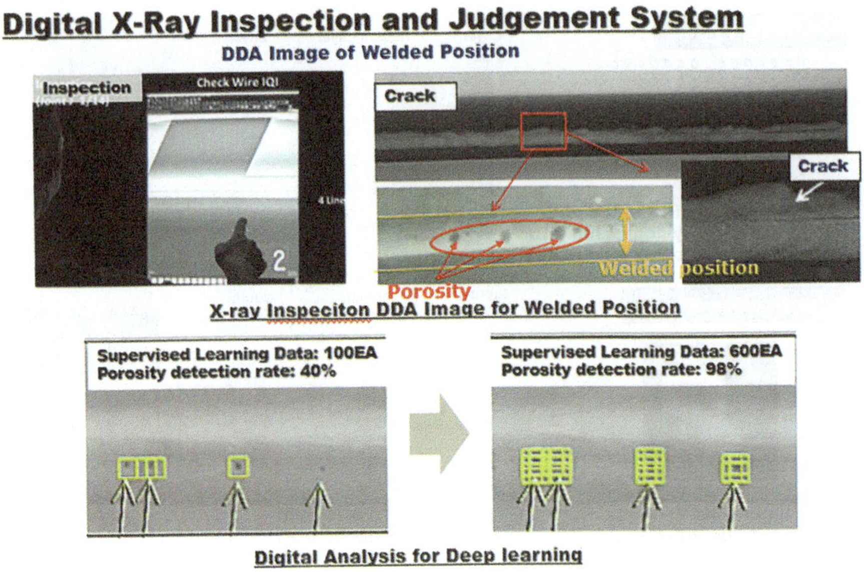

Fig. 13 Analysis results for deep learning of automated digital X-ray judgement system

Figure 13 shows the digitized X-ray inspection image for and artificial intelligence judgement to some defects in image. As shown in this image, the cracks and porosities are shown in the weld metal. When the number of supervised learning data is 100EA, porosity detection rate is only 40%. When the number of supervised learning data is 600EA, porosity detection rate increased to 98%. It has been concluded that we could judge the defects on EBW welded position using machine learning method.

As a result of the development of above-mentioned technology and its application to actual production processes, the time used by inspectors for judgment has been reduced by 80%. In addition, the use of digital data instead of film has eliminated the cost for film and storage, reducing the total cost by 60%.

6 LOC Optimization of Work in Process for Manufacturing Process

In the production flow for aircraft parts, there are some welding related parts like electron beam welding, X-ray inspection and some finishing process. Figure 14 shows the Logistics Operation Curves (LOC) calculation method. This is a method for both increasing throughput and reducing lead time by analyzing the relationship between throughput and lead time of processed products. This method's feature, less burden of model creation during simulation, easier interpretation of results compared to the queueing theory. Figure 15 shows the calculation function for the

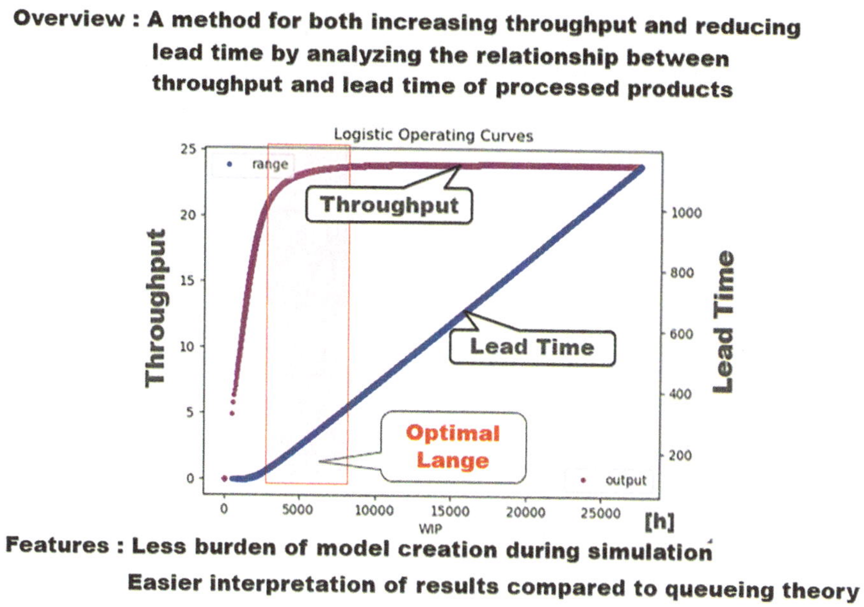

Fig. 14 Relationship between throughput, WIP (work in process) and logistics operation curves

From the theory, the relationship between WIP, throughput, and lead time in a typical job shop workplace can be formulated as follows

Throughput	$OUTPUT(t) = OUTPUT_{max} \times \left(1-(1-t^{c})^{\frac{1}{c}}\right)$
WIP	$WIP\,(t) = WIPI \times \left(1-(1-t^{c})^{\frac{1}{c}}\right) + WIPI \times \alpha 1 \times t$
Lead Time	$Range(t) = \dfrac{WIP(t)}{OUTPUT(t)}$

$WIPI = \sum WC^2 \;/ \sum WC$: *Minimum WIP when ideally operated*

WC : *Amount of work for each workpiece [h]*

$OUTPUT_{max} = \min$ *Available equipment operating hours and worker working hours*

$\alpha 1(=10), c\,(=1/4)$: *coefficient* t : *coefficient of mediation*

Fig. 15 Calculation method and functions for logistics operation curves

logistics operation curves. Throughput, WIP (Work in Process) and lead time are represented by exponents transformed as some dimensionless numbers. As a result, this method can calculate the optimal WIP range from throughput and Lead time information.

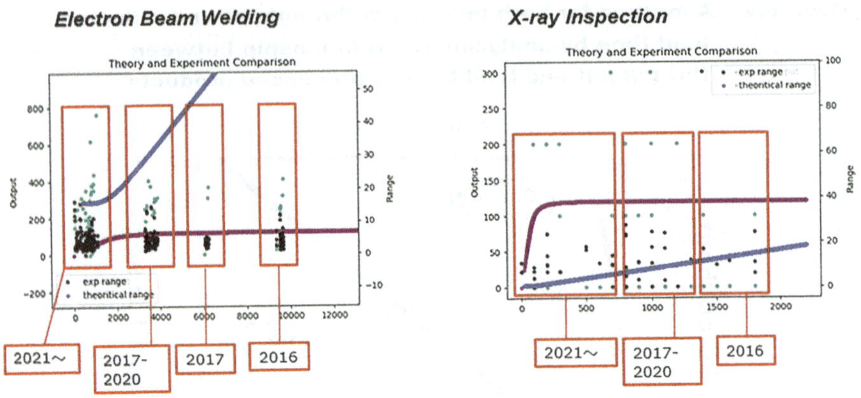

Fig. 16 LOC calculation results of electron beam welding process and X-ray inspection process

Figure 16 shows the calculation results of electron beam welding process and X-ray inspection process since 2016 to 2021 of each process. Electron beam welding process had bottlenecks in 2016–2017. In order to get within the optimum range, improvements were made by reviewing the electron beam welding conditions and systematizing the welding process, resulting in shorter lead times and increased throughput. As a result, the electron beam welding process is able to produce in the optimal range in 2021. In X-ray inspection process, there are bottlenecks in 2016–2020. In the development of an automated system to ensure the optimal range is reached was the result of the development of an automated system, it allows for production in the optimal range in 2021..

7 Optimization of Production Sequence Using Dispatching Order

Figure 17 shows the dispatching order calculation method for machine action sequence. In this method, optimal solution calculation logic for automatic planning and execution instructions as called "dispatching".

This is the relationship between parameter (dispatching order) and evaluation value. Multiple initial values are given using a parallel processing dispatching function calculation. The multiple local search solutions as candidate solutions are derived through repeated calculations to obtain the optimal solution. This method allows to search for the solution with the highest throughput and the shortest time in the process sequence in the manufacture of aircraft components and in the order of operations within multiple processes.

Figure 18 shows the 2D and 3D simulation results of the cleaning process prior to the electron beam welding process using the dispatching order calculation method. This cleaning process is an automated process using a crane and robot. Hot

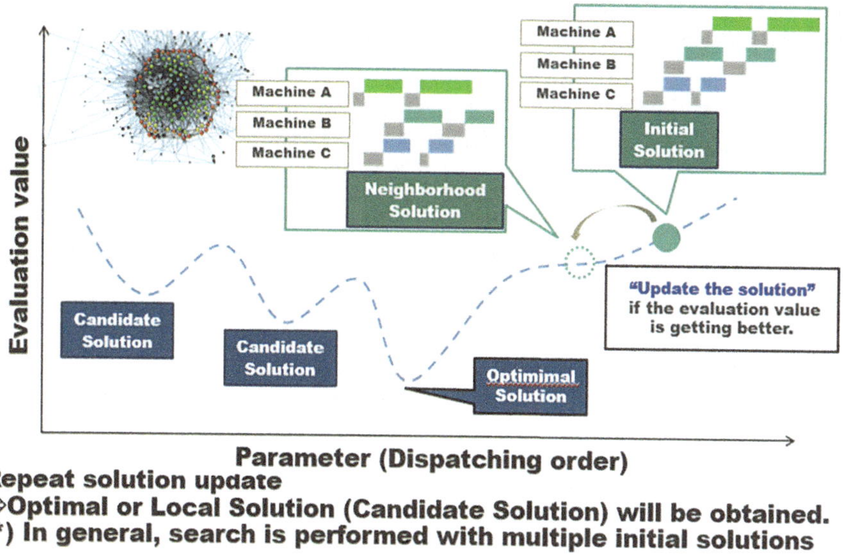

Fig. 17 Dispatching order calculation method for machine action sequence

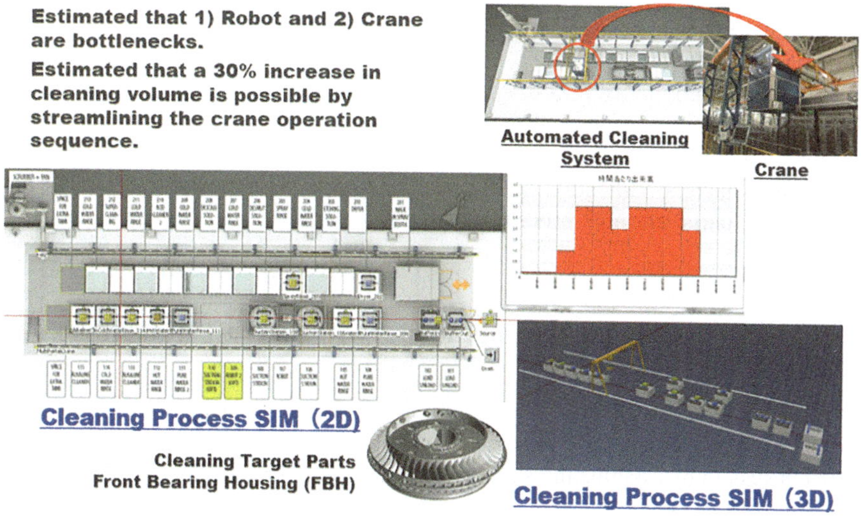

Fig. 18 Simulation results of automatic cleaning process prior to the electron beam welding process

water, alkaline, and acid cleaning baths are available, and robots dip and pull parts in these baths in a predetermined sequence. The calculations would need to be made for the cleaning sequence of each part by robot and crane. The cleaning time required for each part is set in advance, and the set time is input into the sequence

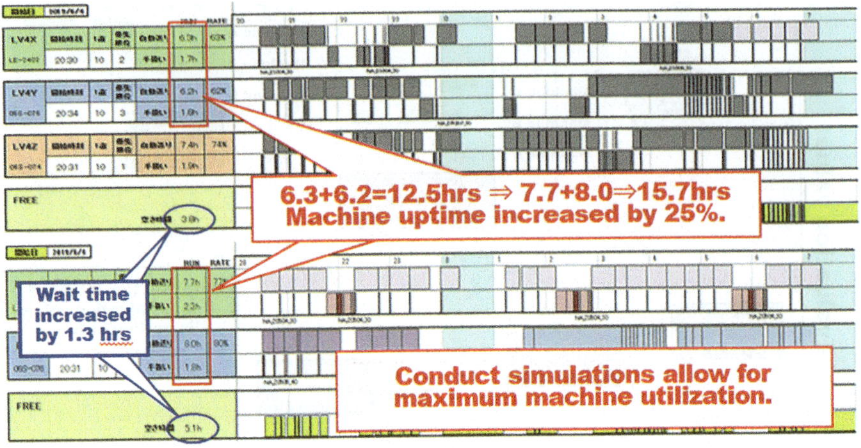

Fig. 19 Dispatching order calculation results for the optimized sequence for automated cleaning

setting by dispatching calculations. Boundary conditions are set so that the waiting time before each cleaning operation is minimized. It was determined that optimizing the cleaning sequence would increase throughput by 30% and reduce lead time by 20% using the dispatching order calculation.

Figure 19 shows the calculation research results for the optimized sequence for automated machining equipment. Before optimal calculation, the planned machining equipment uptime is 12.5 h. After optima calculation, the planned machining equipment uptime is increased to 15.7 h. As a result, the machining equipment uptime is increased by 25%, and this optimal calculation and simulation allow for maximum machine utilization and can be applied to the production sequence for aero engine parts' cleaning process.

8 Strategy for Digital Smart Factory in Aircraft Industry

The artificial intelligence developments in the aircraft industry production were described. Based on these achievements, it is necessary to realize an integrated production system by establishing strategy for digital smart factory in aircraft industry.

Figure 20 shows the engineering steps to smart stage at integrated manufacturing system. In Stage 1, the highly skilled operator knowledge and equipment failure prediction data are converted into digital data. Next, it will be optimized that the simulation of advanced welding and machining process and developed technology that can contribute to automation and productivity improvement by converting the logic used by skilled operator in Stage 2. The scope of automation of welding and machining programs is expanded in Stage 3. The feedback of inspection and evaluation results to processing programs is also expanded, and digitized base data is laterally linked in this stage. It will be able to establish the manufacturing

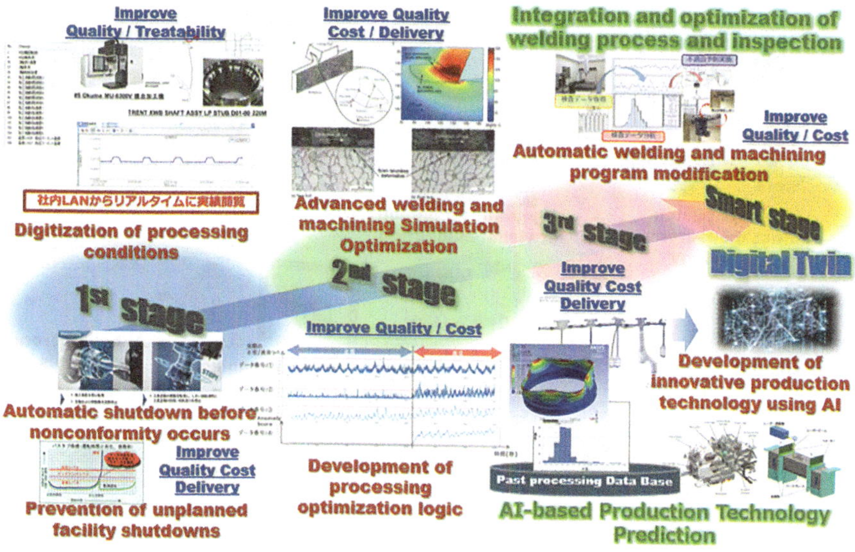

Fig. 20 Engineering strategy of welding process for smart stage

engineering that leads to innovative production technology for the purpose of improving engine performance and productivity by using artificial intelligence technology in smart stage.

The engineering steps of inspection engineering for smart stage are shown in Fig. 21. In this engineering strategy, the system construction in which advanced non-contact inspection technology and inspection measurement data management technology are developed in Stage 1. The development phase and implement judgment technologies are developed in Stage 2. The expansion of these technologies and data for optimization of a wide range of production systems is conducted in Stage 3. 3D superimposition and augmented reality technologies to evaluate complex shaped parts can be used in Smart Stage. In addition, artificial intelligence will be able to analyze all past data in this stage.

The technology strategy stages in the field of production control technology are shown in Fig. 22. Stage 1 is the stage of development and application of system construction technology that enables high-frequency and high-density accumulation of digitalized status data on the production status of people, products, and facilities both within and outside the company. Stage 2 is the stage where lead-time reduction is achieved through optimization of each process, which is expanded from the vast amount of status data accumulated in Stage 1, as well as optimization of all processes from the first to the complete production process. Stage 3 will involve a wide-ranging linkage of AI technologies ranging from micro production sequence optimization to macro production resource optimization calculated from capacity measurement logic. After these stages, in the smart stage, optimization technology is deployed throughout the company and extended to the entire external supply

Fig. 21 Engineering strategy of inspection process for smart stage

Fig. 22 Engineering strategy of production control process for smart stage

chain to establish a fully optimized production system that can automatically and immediately respond to various variable factors when they occur.

Subsequently, a system corresponding to the Smart Stage of inspection and evaluation technology will be integrated with the Smart Stage of production technology as a PLM line, and a system corresponding to the Smart Stage of production

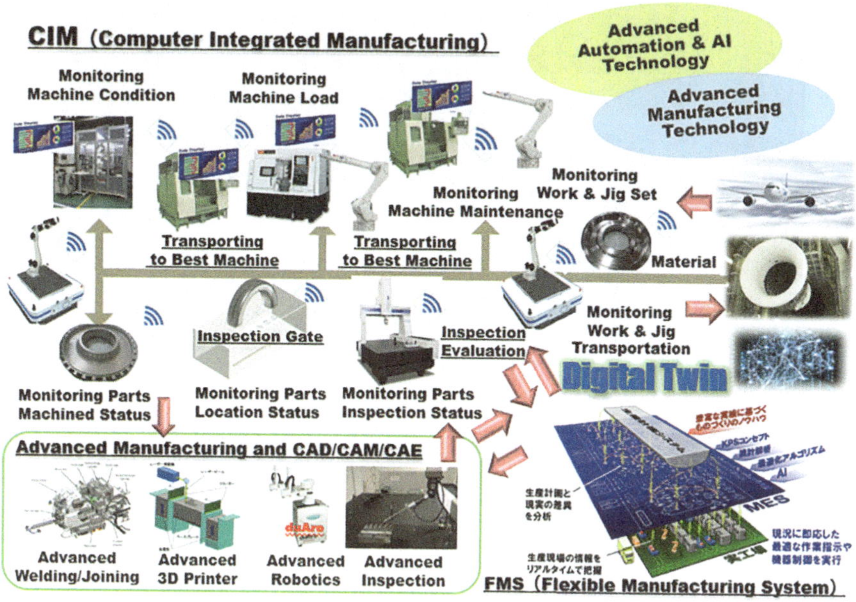

Fig. 23 Schematic illustration of digital smart factory in aircraft industry

management technology will be integrated with the PLM line as a SCM line, thereby integrating the three technology fields and enabling the digital system corresponding to Smart Stage in production technology will be integrated with Smart Stage in production technology as a PLM line.

Figure 23 shows the digital smart factory vision. In this factory system, all information on manufacturing (processing conditions, production capacity, planning, etc.) will be monitored in real time. CPS will reconstruct them as useful digital twin images. Each computer aided engineering system performs a reverse simulation. Meanwhile, advanced simulation and optimization instructions are issued frequently to optimize production conditions.

The welding and inspection technologies are demanded to be further developed while incorporating the advanced digitalization technologies. To this end, "Human to Human: H2H" systems must enter "Human to Machine: H2M" system phase. In addition, the various manufacturing knowledge is standardized through digitalization of data, and artificial intelligence programs will be realized to achieve more precise operations.

The collected data and approximation theory will then be used to develop "Machine to Machine: H2M" system and "Human with Machine: HWM" system across disciplines. Based on this development, IoT and artificial intelligence technology and robotics technology are expected to be integrated and further developed as more advanced intelligent manufacturing systems.

Fig. 24 Hydrogen aircraft image for the future

On the other hand, there is a trend to develop new-concept aircraft that will be a game changer in GX (Green Transformation) such as hydrogen aircraft as shown in Fig. 24 in the aircraft industry. For this development, it is necessary to research and develop new core technologies based on the manufacturing technologies accumulated through DX [4].

9 Conclusions

In this chapter, the machine learning of standardized 3 M data in advanced manufacturing system is described. In this advanced manufacturing system, robot welding system incorporated the skilled operator knowledge was established. Artificial intelligence technology that uses a machine learning method based on multilayer neural network is applied to inspection process. In production control fields, the optimal calculation method classified as artificial intelligence technology allowed to improve the WIP (work in process) range to maximize throughput and minimize lead time by applying automation and production systems. It was also found that the application of a production sequence optimization method using the dispatching method can maximize equipment uptime in the automatic cleaning process prior to welding.

References

1. Tsuzuki, R. (2019). Advanced manufacturing science and technology for innovative production of aero engine -integration of innovative manufacturing engineering and ICT/IoT technology. *Journal of the Japan Welding Society, 88*, 263.
2. Tsuzuki, R. (2021). Digital smart factory initiatives and production improvement in aircraft industry -development of automation and AI technology in welding and inspection process for intelligent production system. *Journal of the Japan Welding Society, 90*, 44.
3. Tsuzuki, R. (2021). Aircraft engine titanium alloys, welding/joining process and automation technology for future. *Journal of the Japan Welding Society, 90*, 234.
4. Tsuzuki, R. (2022). Super heat-resistant materials and welding/joining technologies for aero engines. *Journal of the Japan Welding Society, 91*, 127.

Digital Twin Architecture for a Sustainable Control System in Aircraft Engines

Maryam Farsi, Bernadin Namoano, Christina Latsou, Vaishnav Venkata Subhadu, Haoxuan Deng, Zhen Sun, Bohao Zheng, Davide D'Amico, and John Ahmet Erkoyuncu

1 Introduction

Since the late 1990s, gradual developments in computer science, data analytics, modelling, and simulation techniques reduced the dependencies on traditional, time-consuming, and costly physical testing equipment for complex engineering assets [1]. High-value industries historically prioritised commercial gains and performance-driven asset value creation. However, over the last decade, they have been impacted by environmental concerns and several obligations to reduce their emissions and carbon footprint. The application of modelling and simulation techniques in the aviation industry have been studied to enhance fuel efficiency with the aim of enhancing thermal cycle efficiency [2–4]. The aviation industry is projected to experience continuous growth, particularly in passenger air traffic, at an annual rate of 4.7% [5]. This growth is anticipated to contribute to global environmental concerns at national and international levels, including climate change and CO_2 emissions. Despite improvements in fuel efficiency, the aviation industry is facing a critical environmental crisis, currently contributing to 2.5% of global CO_2 emissions (i.e. roughly 1 billion tons of CO_2 per year) [6]. The industry has set ambitious goals to reduce emissions to half of 2005 levels by 2050 and improve fuel efficiency by investing in green alternative fuels [7]. The requirement for continuous efficiency improvements relies on technological strategies such as aircraft weight reduction (using lighter and more wear-resistant materials), engine propulsion

M. Farsi (✉) · B. Namoano · C. Latsou · V. V. Subhadu · H. Deng · Z. Sun · B. Zheng · D. D'Amico · J. A. Erkoyuncu
School of Aerospace, Transport and Manufacturing, Cranfield University, Cranfield, UK
e-mail: Maryam.Farsi@cranfield.ac.uk; Bernadin.Namoano@cranfield.ac.uk; Christina.Latsou@cranfield.ac.uk; V.VenkataSubhadu@cranfield.ac.uk; Haoxuan.Deng@cranfield.ac.uk; Sun.Zhen@cranfield.ac.uk; Bohao.Zheng@cranfield.ac.uk; R.DAmico@cranfield.ac.uk; j.a.erkoyuncu@Cranfield.ac.uk

T. H. Karakoc et al. (eds.), *Sustainable Materials and Manufacturing Techniques in Aviation*, Sustainable Aviation, https://doi.org/10.1007/978-3-031-62987-7_5

(thermodynamic and propulsive) efficiency, and aerodynamic (drag reduction) improvements. The adoption of aerodynamic strategies to reduce the environmental impact of the aerospace industry optimises state-of-the-art aerodynamic technologies and tools for improving aerodynamic efficiency, devices that have an inherent potential to improve efficiency [8].

1.1 Sustainable Aviation Industry

Sustainability is seen as being at the core of business strategy, as highlighted at the United Nations Conference on Sustainable Development 2030, which includes smart manufacturing, energy-efficient buildings, and low-impact industrialisation [9]. In the aviation industry, fuel efficiency and Sustainable Aviation Fuel (SAF) are the most critical approaches towards decarbonisation and sustainable aviation. SAF is made from waste and other residue fats, oils, and gases and produced through the hydro-processing of esters fatty acids known as HEFA. Currently, most certified SAFs can be blended with conventional fossil jet fuel up to 50% [10]. Modelling the behaviour of energy consumption is the first step in monitoring and reducing energy consumption. However, it requires an in-depth understanding of the motion and dynamic behaviour of the system [11]. Besides, electrification of the engine fuel pumping system is a priority issue due to the contribution of the engine's power generation structure to the reduction of fuel consumption. A multi-physical field coupled model is required to numerically simulate the heat dissipation characteristics of the housing under typical operating conditions. On the one hand, the heat generated by motor losses should be reduced. On the other hand, the heat dissipation capacity of the motor should be improved as much as possible from the perspective of cooling performance [12]. Decarbonisation and reducing environmental impact are necessary for a sustainable future in the industry. The key approaches towards sustainable aviation are presented in Fig. 1.

The developments in modelling and simulation approaches enable a more accurate representation of systems as well as a prediction of the systems' behaviour. These led to the idea of connecting the virtual space to the real world through the concept of 'Digital Twin' [13]. The rapid development of simulation technology has pushed the boundaries from purely model-based numerical simulations such as large eddy simulations (LES) and Reynolds-Averaged Navier Stokes (RANS) to hybrid modelling utilising experimental data and physical laws [14]. Data-driven methods are a new modelling paradigm that addresses the problem in complex system and process modelling since it can directly create the mapping between engine input and output using collected or streaming data without explicit model construction. It has shown great potential in statics analysis for fluid dynamics, engine system identification, fuel efficiency, and health monitoring because of its capability for high-dimensional non-linear function approximation and general data

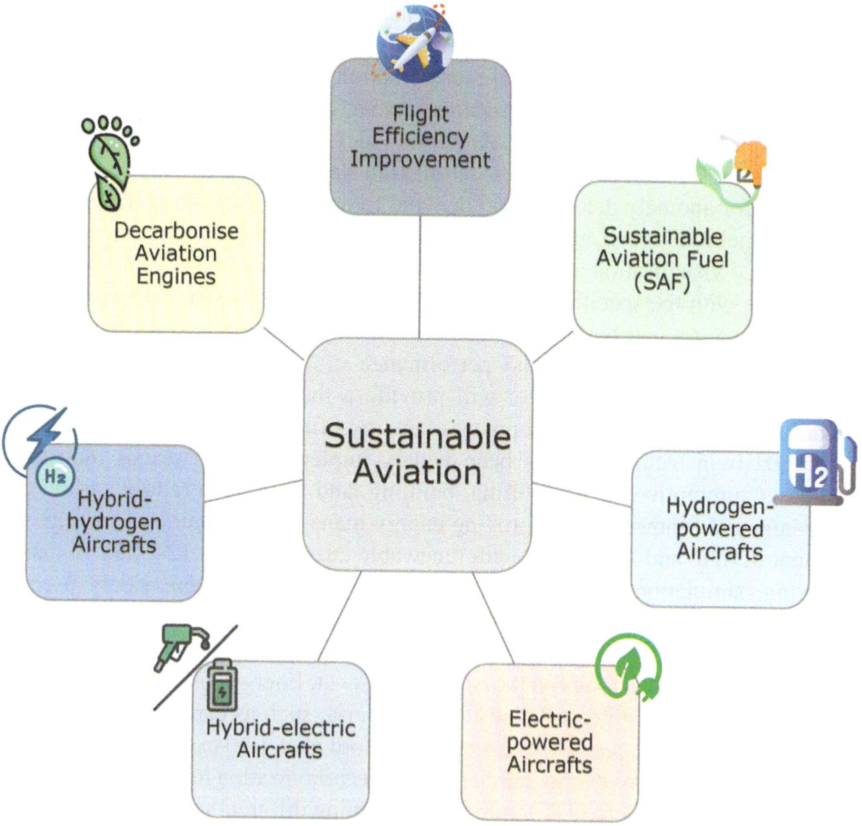

Fig. 1 Sustainable aviation

interpolation [15]. Later, with the great leap in sensing technology like the Internet of Things (IoT), this concept of bridging physical sensors with actuators and digital decision-making modules with controllers is expected to be realised for an industrial application [16, 17].

1.2 Digital Twin Technologies and Sustainability

Digital Twin (DT) technologies are seen as a major contributor to automation, digital manufacturing, and sustainability. Grieves [18] defined DT with three elements: a physical product in the real space, a high-fidelity virtual product in the virtual space, and a connection that ties these two spaces together. In the scientific literature, there are several definitions of DT. Recently, ISO 23247-1:2021 defined DT in the manufacturing domain as a fit-for-purpose digital representation of an observable manufacturing element with synchronisation between the element and its digital representation. However, to include a broader range of applications, a DT can be defined as a dynamic digital representation of a real entity [19]. During the Industry 4.0 era, researchers and practitioners are seeking a flexible and intelligent solution for smart energy systems. Ardebili et al. [20] suggested that DT can make energy systems intelligent in decision-making and provide dynamic monitoring capability for anomaly detection and demand forecasting [20]. The DT models can provide a method for self-monitoring sustainable development [21]. Moreover, DTs can enhance visualisation and allow end users to select the appropriate strategy in accordance with the specified sustainability goals [22]. Combined simulation platforms for engines and high-value asset equipment can be used to assess and optimise the energy distribution and performance of future vehicles with different technologies or strategies. These will provide a theoretical basis and numerical modelling support for the development of energy-efficient technologies [23].

Digital twin technology has been widely implemented in various industries, including automotive, manufacturing, building, and aircraft, to reduce carbon and environmental footprints by improving energy management, maintenance, energy-efficient design, and integration with renewable energy sources [24, 25]. In manufacturing, simulation modelling and DT technology have been widely used to analyse and improve system performance [26]. Franciosi et al. [27] observed increasing interest in using DT technology for sustainable industrial maintenance and production, with a focus on the economic aspect. Energy cost and efficiency are commonly studied, while pollutant-related criteria, such as carbon/CO_2 emissions and waste generation, receive less attention. Alford et al. [28] explored the potential of flexible DTs to drive energy efficiency and decarbonisation in process industries. They introduced a layered DT approach, combining different tools to create a self-optimising and self-learning virtual plant. Combining energy-efficient technologies with smart and digital technologies can solve the problem of energy consumption [29]. Li et al. [23] suggested that for high-value assets such as engines, combined DT and simulation platforms can be used to assess and optimise energy distribution. Such an approach can provide a theoretical basis and numerical modelling support systems for the development of energy-efficient technologies [23].

Digital twin-driven product design helps to close the gap between the product's physical and virtual spaces, which decreases the efficiency and sustainability of the design, manufacturing, and service. Web-based digital twin architectures have enormous potential to contribute to the sustainability of industrial cyber-physical systems. Data-sharing and collaboration enabled through the DT technology create opportunities for companies to improve sustainability performance, even on legacy assets [30]. Ezhilarasu et al. [31] concluded that in the current climate, the main problem with the management approach to aircraft health is the inflated cost of efficiency. They believed that the cascading effect of aircraft would be mitigated by building DT technologies for aircraft and creating a unified system architecture for isolating faults [31].

1.3 Contribution to Knowledge

The DT model of a sustainable control system in aircraft engines can be described as a digital copy of the control system in a virtual environment that enables (near) real-time monitoring and restricting engine emission (CO, NOx, etc.) and system running state followed by decision automation and system adaptive control [32]. Such a DT model allows the control system to be self-aware of its current state and adapt to external changes dynamically. A schematic model of an aircraft engine, the critical modules and the allocation of the current control system is presented in Fig. 2. The critical modules include the High Pressure (HP), Intermediate Pressure (IP) and Low Pressure (LP) sub-systems. The primary role of the engine control system is to provide the suitable amount of fuel needed for the engine to produce the desired power based on the pilot's power request and maintain the engine power at the desired level in the presence of airflow disturbance and changes in flight conditions. It is evident that ensuring an engine's sustainability and efficiency relies on a wide range of data and information encompassing the operational environment, engine's specifications and health condition. This requirement presents the prevalent challenge of inadequate data availability and quality to support the sustainability and efficiency targets in the aviation industry. This paper suggests the application of DT technology as a promising approach to overcome this challenge. This vision is supported by a systematic literature review and experts' insight in this paper.

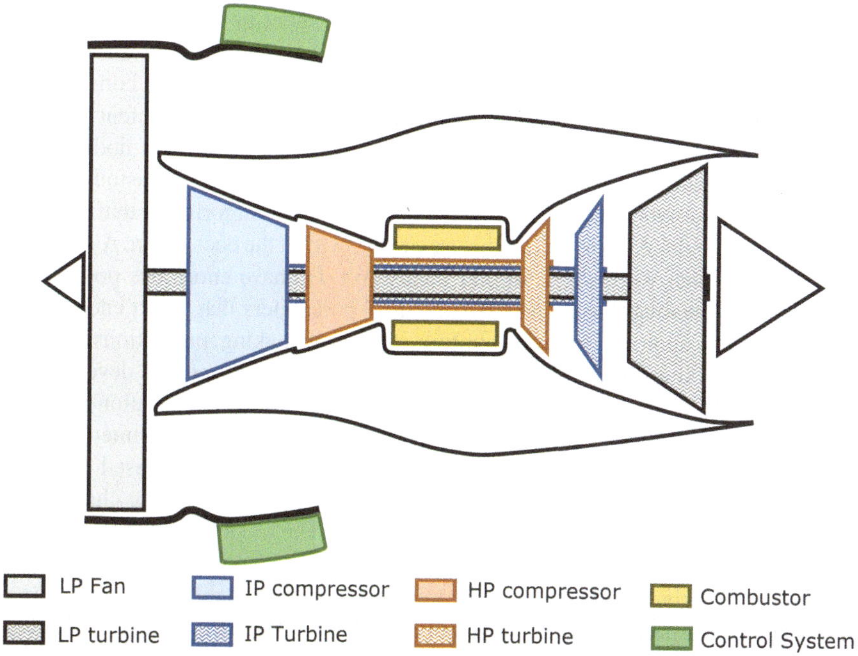

Fig. 2 Aircraft engine and its critical components

In this paper, the DT architecture of a sustainable control system for an aircraft engine is presented. This study attempted to answer the following research question: How can DT assist with improving engine sustainability and efficiency in the aviation industry? The paper is organised as follows: Section 2 presents the state-of-the-art regarding DT designs, sustainability, and fuel efficiency in complex engineering assets using a systematic review approach. The research methodology and the overview of the analysis of the methods used in this research to develop the DT architecture are presented in Sect. 3. The proposed digital twin architecture is presented in Sect. 4. This is followed by discussions in Sect. 5 and conclusions and further work in Sect. 6.

2 State-of-the-Art

2.1 Digital Twin Technologies

Digital twin (DT) is known as one of the emerging technologies within the industry and construction sectors as an effective approach for improving energy efficiency and management [33]. The use of DT and Artificial Intelligence (AI) technologies in energy consumption and efficiency have been studied recently, focusing on effective energy management for buildings [34, 35] and electric vehicles [36], energy efficiency in hydraulic systems [37], energy efficiency and maintenance in smart manufacturing systems, and carbon footprint assessment in smart cities [38]. DT technology is endowed with a value system that enables it to make autonomous decisions. Such capability can be used to promote sustainable energy consumption in ecosystems [39]. Barykin et al. [40] introduced an energy-efficient network topology for digital omnichannel marketing. The network consists of nodes representing participants with unique characteristics, creating a digital customer value network. The authors developed omnichannel interaction algorithms that incorporate economic indicators, enabling the use of DTs within the ecosystem. Assad et al. [41] proposed that Web-Based Digital Twins (WDTs) have enormous potential to contribute to sustainability by accessing control parameters that affect energy consumption, recording energy consumption data, and making predictions through computational algorithms [41]. Later, Zhao et al. [42] analysed the development modes of a DT system, including symbiosis, evolution, and optimisation and proposed the key technologies of a DT for an integrated energy management system [42]. In addition, Sun et al. [43] developed a deep learning model based on graph neural networks (GNN) to predict physical states and thermodynamic characteristics in supercritical CO_2 power generation systems. The model outperformed traditional machine learning models.

In manufacturing, the DT-driven optimisation approach enables the smart manufacturing of aerospace generators with a high number of system components through the virtual-real combination of system performance and manufacturing. In such applications, the DT model receives real-time data from manufacturing measurements and feedback to the key performance indicators, which are corrected throughout the optimisation process [44]. Moreover, DT can be applied to optimise energy efficiency in manufacturing systems. In that case, data can be collected in real-time from analogue physical systems and transferred to digital systems. Then, energy-saving strategies and key monitoring metrics can be realised in the digital space. Afterwards, he generated energy-saving instructions that can be sent to the analogue manufacturing system in the physical space and control the energy consumption [45]. Zhang et al. [13] proposed a framework for equipment energy consumption management in a digital twin shop floor to reduce energy consumption and improve the energy efficiency of equipment on the shop floor. They argued that the digital twin shop-floor environment could be used to explore the potential applications of equipment energy consumption management under three categories: energy consumption monitoring, energy consumption analysis, and energy consumption optimisation. Li et al. [46] developed a data-driven DT model for monitoring and controlling the energy behaviour of equipment in manufacturing. The model combines Petri-net modelling and machine learning techniques and utilises real-time data on working conditions, operating parameters, and production load to enhance the energy management capacity of the DT. In a more recent study, Krommes and Tomaschko [47] proposed a DT framework for value and material flow when monitoring energy and material flows. The framework proposed solutions for allocating energy consumption and setting peak energy savings.

2.2 *Digital Twin Applications*

The application of DT technologies for energy efficiency of data centres has been investigated by several researchers. Tang et al. [45] highlighted that the implementation of DT technology is crucial for achieving the standardisation and modularisation of industrial data infrastructures, resulting in significant energy savings for data centres. By adopting DT, traditional industries can transform into energy-efficient sectors aligned with the advancements of Industry 4.0. This transition not only enhances industrial energy efficiency but also extends the scope of innovative technologies in this domain [45]. Similarly, Dongyun et al. [48] proposed an intelligent online operation and maintenance system based on DT and industrial internet technologies for real-time monitoring and management of equipment and the operational status of hydrogen energy production and storage power plants. Their proposed architecture considers the virtual model of a physical power plant, visual display, status evaluation, information viewing, traceability, optimisation, and statistical analysis. Moreover, Zinting et al. [49] proposed an energy-saving solution using Artificial Intelligence (AI) and DT for a smart data centre. Their work concluded that the DT model can reduce energy consumption by optimising the cooling system.

The application of DT technology for evaluating sustainability and energy consumption in high-value and complex engineering assets has been studied recently. It is crucial for the DT models to access the control parameters affecting energy consumption and to record the energy consumption [41]. Passath et al. [50] present the application of DT technology for combined critical value assessment and RAMS2 (Reliability, Availability, Maintainability, Safety,and Sustainability) over assets' life cycle. Deon et al. [51] proposed a DT architecture based on a decision support system for real-time predictive maintenance and fault classification in thermal generation engines. The architecture integrated machine learning models and fault classification algorithms into the plant's control system, enabling operators to receive timely alerts and insights. A DT methodology using Computational Fluid Dynamics (CFD) analysis is proposed by Gebauer et al. [52] to optimise the cooling of induction motors. The proposed DT minimised losses and maximised efficiency by considering complex geometry and asymmetric flow caused by the rotating fan. Several studies focused on the application of modelling and simulation to predict the performance of critical components in a gas turbine, in particular [53, 54]. Xin et al. [55] focused on the application of DT technologies for the design of gas turbines. They suggested that the DT model can improve the efficiency of design data transmission and establish a large database for test evaluation. Sciatti et al. [56] developed a Simulink-based simulation model for the fuel system in aircraft gas turbine engines. The system's performance was evaluated under different conditions, considering components like tanks, pumps, valves, and fuel metering units. The simulation results and power consumption analysis demonstrated the model's effectiveness in assessing various operating conditions and component geometries.

For sustainable aircraft engines, the application of DT focused on fuel systems. A DT model of a fuel cell for a hybrid electric vehicle is proposed by Bartolucci et al. [36]. Their proposed model simulates the dynamic behaviour of hydrogen energy systems and auxiliary systems with high fidelity. They used fuzzy logic controllers to implement energy management strategies and optimised control parameters through genetic algorithms to adapt to different driving conditions. Stoumpos et al. [57] explore the safety issues of marine dual fuel (DF) engines under different operating modes, especially considering the failure or failure of sensors and actuators. The authors used GT-ISE software to construct a DT model of DF engines, which can simulate the engine's response under steady-state and transient conditions. They also used Fault Operation Simulator (FOS) to simulate faults or failures of sensors and actuators. The author used the OREDA database and other databases to obtain the failure rate and failure operation data of engine components. A WEG Digital Twin application composed of motors and controls in the WEGnology Internet of Things (IoT) platform is proposed by Junckes et al. [58]. The proposed model provides an increase in energy efficiency and generates insights for the design of new products. Duarte et al. [59] proposed a DT model for predicting pressure in fuel injection systems using machine learning. In their study, the Random Forest algorithm has been used to implement the DT model to predict the pressure level in the fuel injection system. In a different study, Zhu et al. [60] proposed a DT model for vibration monitoring and control in maglev motors.

2.3 Research Gap

In the existing DT technologies, efficient information transfer is incredibly challenging due to resource constraints, random tasks, and resource heterogeneity. In this process, DT technologies should be able to efficiently allocate dispersed resources in the network and reduce energy consumption while improving data processing efficiency [61]. DT technologies can improve efficiency and automate complex engineering assets. However, they require enormous amounts of real-time and historical data for effective simulation and prediction. Aliramezani et al. [62] reviewed the research around the modelling and optimisation of internal combustion engines and highlighted the existing challenges that can be resolved by machine learning techniques, including supervised learning, unsupervised learning, and reinforcement learning. In their study, the identified challenges are real driving emission modelling and control, combustion knocks detection and control, combustion mode transition in multi-mode engines, combustion noise modelling and control, combustion instability and cyclic variability control, costly and time-consuming engine calibration, and fault diagnostics of some ICE components [62]. In the context of an aero-engine, applying the data-driven methods for extracting the relationship between the engine inputs and outputs for direct thrust control is mentioned in [63]. Mouzakitis et al. [64] suggested that AI models with high-performance computing systems can support DT models to run smoothly [64].

Agouzzal and Abbou [65] argued that DT technologies are key for achieving low-cost and sustainable energy transformation, but there are also some technical, standard, and security issues that need to be addressed. Computation of enormous amounts of data with limited resources is a major obstacle to improving the Industrial Internet of Things (IIoT) service level. To reduce the communication overhead between the physical devices and the digital space, a federated learning approach can improve communication efficiency and reduce transmission energy consumption. The combination of the DT model and asynchronous transmission scheme to solve the sub-problems based on the deep neural network model to further decompose the problem is one of the approaches [66]. Bortolini et al. [67] reviewed the applications of DTs in energy-efficient buildings. They acknowledged the potential of DTs while highlighting challenges in data integration, decision-making, sensor selection, and data visualisation. Artificial neural networks (ANN) and hybrid approaches combining white-box and black-box models are commonly used for developing DTs in this context.

The existing literature on DT, system efficiency, and energy consumption in different applications confirms that DT can be a promising technology to monitor and control systems' emissions, energy consumption, and efficiency. This paper contributes to the existing literature by expanding the application of DT in the aviation industry and proposing the DT architecture for a sustainable control system of an aircraft.

3 Methodology

The overall research methodology to carry out this work is presented in Fig. 3. The proposed methodology includes five main steps: theoretical background, problem statement, DT architecture development, evaluation of the DT architecture and finally, discussion and impact analysis.

3.1 Theoretical Background

The research work is mainly built on a systematic review of the existing literature in the aviation industry, sustainability, digital twin, and energy and fuel efficiency. To establish the state-of-the-art within the scope of the research, a two-folded systematic literature search is carried out. Scopus was used to find the relevant documents. Besides, the existing frameworks, architectures around DT and the relevant Standards are searched through Google Scholar. In the Scopus repository search, no limitations are applied in terms of year, source title, source type, and keywords. However, the documents on healthcare, medicine, agriculture, social science art, and humanities are excluded. Moreover, books, notes, and abstract reports are also excluded from the document type. Documents in English and Chinese languages are included since some of the co-authors knew the Chinese language. Two sets of keyword strings are used to create the initial database of documents in Scopus.

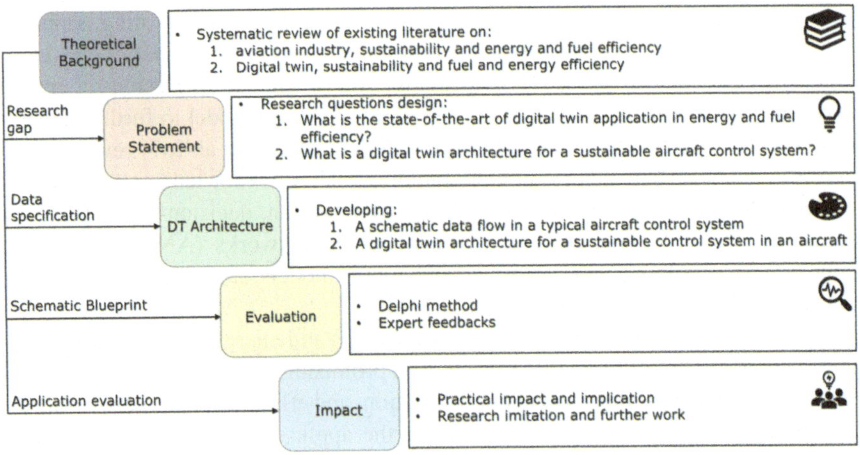

Fig. 3 Research methodology

1.	TITLE-ABS-KEY (aircraft OR aviation OR aerospace OR aeroplane) AND TITLE-ABS-KEY (manufactur*) AND TITLE-ABS-KEY (efficien* OR sustainab* OR resilien* OR productivity OR agil* OR lean) AND TITLE-ABS-KEY (energy OR fuel) AND TITLE-ABS-KEY (engine)
2.	TITLE-ABS-KEY ("digital twin"OR "digital thread" OR "virtual twin" OR "cyber twin" OR "digital replica" OR "cognitive twin" OR "cognitive digital twin"OR "digital data hub" OR "digital shadow" OR "digital mirror") AND TITLE-ABS-KEY (efficien* OR sustainab* OR resilien* OR productivity OR agil* OR lean) AND TITLE-ABS-KEY (fuel OR energy)

The initial search yielded 1363 documents, including 574 and 789 from the first and second keyword strings, respectively. In the documents' screening stage, the title and abstracts of documents were reviewed to shortlist the documents relevant to digital twin applications in the sustainable aviation industry and complex engineering assets. At this stage, duplicates and conference reviews are excluded. Also, the documents on the topics of sustainability in DT, efficiency in building DT, integrated energy systems in housing and construction, and network DT are excluded. This resulted in the exclusion of 939 documents. This left a database comprising 212 articles for full-text screening for eligibility. At this stage, the documents without actual contributions, real DT model development, being related to manufacturing in the aviation industry or high-value assets are excluded. This left a database comprising 75 articles for literature review. The literature search completed is based on PRISMA guidelines, as illustrated in Fig. 4.

3.2 *Problem Statement*

A thorough literature review is completed to find the answer to the problem of 'How can DT assist with improving engine sustainability and efficiency in the aviation industry?'. To address this, two main research questions are designed to understand the state-of-the-art digital twin application in energy and fuel efficiency, and the

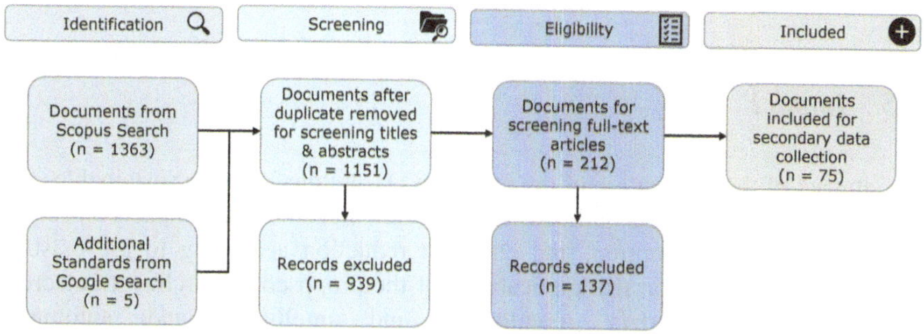

Fig. 4 PRISMA flow chart for collection of documents for literature review

existing knowledge about digital twin architecture for a sustainable aircraft control system. After completing the literature review, the contribution of this work is drawn towards proposing a DT architecture for a sustainable control system of an aircraft.

3.3 DT Architecture

DT architecture should enable effective data management automation as well as online processing and decision-making. Investigations on multiple software, data platforms, and standards models fuelled the present architecture development. ISO 23247 DT reference architecture depends on the standards and technologies available to model the observable manufacturing elements. Different manufacturing domains can use different data standards. As a framework, this document does not prescribe specific data formats and communication protocols. The scopes of the four parts of this series are as follows: (i) ISO 23247-1: General principles and requirements for developing digital twins in manufacturing; (ii) ISO 23247-2: Reference architecture with functional views; (iii) ISO 23247-3: List of basic information attributes for the observable manufacturing elements; (iv) ISO 23247-4: Technical requirements for information exchange between entities within the reference architecture. The developed architecture makes use of the ISO 23247-2 and ISO 23247-4 standards to implement a high-level architecture for the implementation of aircraft engine sustainable control systems.

3.4 Evaluation

Several internal workshops with academic experts in DT, IoT, and data architecture are completed to review, enhance, and verify the proposed DT architecture. The final proposed architecture is presented and discussed in an academic panel at the University Research Centre, and the evaluation is completed using the Delphi method.

3.5 Impact

The impact of the application of the proposed DT architecture in a sustainable control system of an aircraft is outlined. This research argued the DT impacts on energy efficiency, fuel consumption, and emission reduction according to the existing knowledge and based on the applicability of the proposed DT architecture across data management, data monitoring, and intelligent and automated decision-making.

4 Digital Twin Architecture for an Aircraft Engine Sustainable Control System

In this section, a schematic architecture for a typical aircraft engine control system is presented in Fig. 5. This architecture provides an overview of how the control system collects and processes the environmental and operational data in the engine to set the specifications that allow the engine to operate efficiently. The proposed Digital Twin architecture of a sustainable control system for aircraft engines is presented in Fig. 6. The DT architecture extends the capability of the control system by allowing continuous real-time collection, assessment, analysis, and prediction of the fuel and energy consumption of the aircraft to provide a sustainable and efficient operation.

4.1 Aircraft Engine Control System

Before a flight, the request for a change in thrust is initiated by the pilot through the throttle lever [68, 69]. The Power Lever Angle (PLA) captures the change in the angle of the throttle lever and transmits a reference signal to the engine control unit (ECU). The ECU outputs the actual fuel flowrate command signal to the fuel metering valve actuator, which controls the fuel flow rate to the engine combustor, to achieve the requested change in thrust within the operational and safety limits of the aircraft engine. The fan speed (N1) or engine pressure ratio (EPR) can be used to measure thrust through the engine control system using sensors. N1 is the speed of rotation of the LP shaft, and EPR is the total pressure ratio between the inlet and outlet of the aircraft engine. These parameters are fundamental indicators of engine performance, used to control the engine's thrust output. The mapping of thrust into the equivalent N1 or EPR is performed at the setpoint. The difference between the

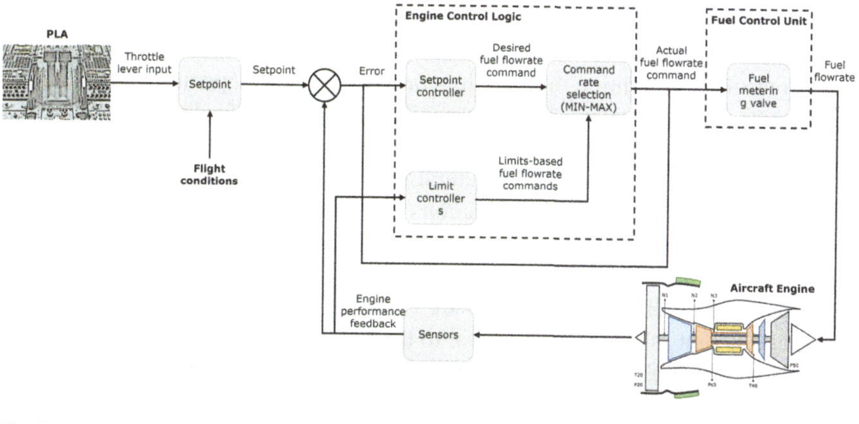

Fig. 5 Typical aircraft engine control system

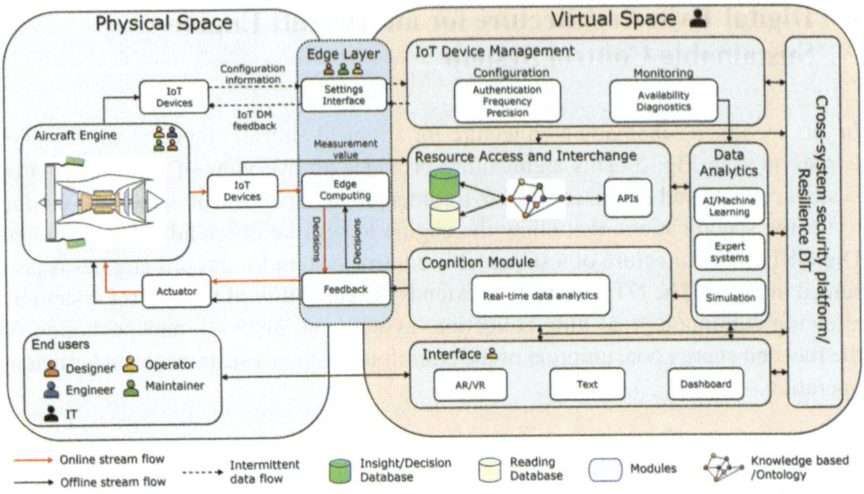

Fig. 6 DT architecture for a sustainable aircraft engine control system

desired N1 or EPR and the actual N1 or EPR is obtained using the sensor data from the engine performance feedback signal. The difference is defined as an error that the ECU reduces while considering the limits of the aircraft engine. In addition, the flight conditions are considered to account for the variations (e.g. variations in air density affect thrust produced) in environmental conditions that can affect the performance of the aircraft engine.

The ECU has three key components: a setpoint controller, limit controllers, and a command rate selection (MIN-MAX) logic [70]. The setpoint controller works on the error signal to determine the desired fuel flow rate that corresponds to the change in thrust request from the pilot. However, an aircraft engine has some critical operational and safety limits, such as maximum rotor speed, maximum temperature of the turbine, minimum compressor surge margin, maximum static pressure to prevent combustor pressurisation, and minimum fuel flow to prevent lean blowout [71]. The purpose of the two other components of the ECU is to act as a protection logic that prevents the aircraft engine from exceeding the operational and safety limits. The limit controllers determine the fuel flow rate based on the operating conditions of the aircraft engine obtained from the engine performance feedback signal. These include a combustion blowout regulator that considers Ps3, a fan speed regulator that considers N1 and P20, and structural limit regulators that consider T48, Ps3, N2, and N3. The command rate selection (MIN-MAX) logic compares the limits-based fuel flowrate commands from the limit controllers and the desired fuel flowrate command from the setpoint controller and selects the one (actual fuel flowrate command) that will ensure no limits are exceeded [70]. The actual fuel flow rate command is sent as an actuating signal to the fuel metering unit (FMU) of the fuel control unit (FCU). In the FMU, the fuel metering valve actuator is adjusted to deliver the required fuel flow rate to the combustor of the aircraft engine, where the fuel is burned to drive the turbine and connected compressors. Thrust is thus produced from the resulting airflow through the engine, including both the core and the bypass flow paths.

4.2 DT Architecture

The proposed DT architecture is presented in Fig. 6. The proposed architecture is developed based on the expert's knowledge of academics in DT as primary data, and following the systematic documents search, review, and collecting the secondary data from the existing literature. The proposed DT has physical space, virtual spaces, and an edge layer. The physical space is composed of aircraft engines and users. The virtual space includes IoT device management, resource access and interchange, cognitive module, interface, and cyber security platform. The edge layer interlinks the physical and virtual domains through the interface, measurement value, and feedback.

4.2.1 Physical Space

In the proposed DT architecture, the physical space consists of the aircraft engine equipped with IoT devices and actuators. IoT devices seamlessly collect real-time data from the engine, its control system, and the environment (e.g. sensors measuring rotor speeds, pressure, temperature, etc.). This data is then transmitted to the virtual space through the edge layer, forming the basis for continuous monitoring and analysis. Additionally, the IoT devices enable bi-directional communication with the IoT Device Management (DM) module, allowing intermittent yet seamless data exchange. The setting interface module within the edge layer allows these devices to be configured and managed, ensuring optimal alignment with desired parameters. Moreover, mechanisms are employed to monitor the operation of the IoT devices and address potential failures, ensuring continuous operation. Furthermore, actuators are employed for making responsive adjustments within the engine based on the feedback received from the edge layer. For instance, the fuel metering valve actuator regulates fuel flow rates to prevent sudden thrust fluctuations, ensuring a consistent and steady engine performance. This guides combustion, powering turbines and compressors for the required aircraft thrust, while enhancing aircraft's efficiency and sustainability.

Within the proposed DT architecture of the aircraft engine control system, five user roles contribute to improving the sustainability and efficient operation of aircraft engines. *Designers* employ DT data analytics insights to guide sustainable design decisions by modifying design and/or material parameters, assessing performance under different conditions and identifying areas for improvement. This can enhance system sustainability and efficiency while also establishing a link between data-driven insights and innovative design practices. *Engineers* use DT-driven insights to conduct in-depth performance analyses. By comparing real-world data with the DT's predictions, they can identify deviations, validate models, and improve engines' efficiency and sustainability. Furthermore, *operators* utilise the DT's virtual data from the cognition module for real-time optimisation, monitoring performance metrics, adjusting parameters and promptly responding to anomalies for enhanced sustainability. In parallel, through continuous monitoring and analysis of data from the DT, *maintainers* can predict maintenance needs more accurately. They can anticipate potential issues, plan maintenance schedules, and replace components before they fail, reducing downtime and extending the lifespan of the aircraft engine. Additionally, *IT professionals* enable seamless DT operation through interface design, data security, and technology integration, ensuring a seamless connection between the physical and virtual spaces. All users derive valuable insights from the interface module, which enables well-informed decisions through Augmented Reality (AR), Virtual Reality (VR), text-based interactions, and visualisations. This collaborative interdisciplinary approach emphasises the significance of expertise across diverse fields and insights driven by users, thus enhancing the engine control system's capacity for efficient and sustainable operation.

4.2.2 Virtual Space

IoT Devise Management The data collected in the physical space is configured and monitored through the IoT Device Management section in the virtual space, which has two main primary functions: one is to sense the state of the environment by configuring the complex environmental information around it, and the other is to monitor the condition of the physical entity itself. Under the set interface protocol, the IoT device collects data according to the established rules. By selecting different frequencies, the system's accuracy is constantly adjusted to achieve the function of system environment configuration. The data of the physical entity itself will be presented through the availability, and the diagnosis of its state can realise the function of monitoring the aircraft engine.

Resource Access and Interchange The resource access and interchange module enables information and knowledge management. Measurements collected in the edge layer are stored, linked, and shared through application programming interfaces (APIs). DT Storage systems should enable high resource availability (fast information storage and retrieval), fault tolerance, and scalability. Three different types of databases are used in this framework—the first type stores data from observable assets or process measurements. The second type of database represents both insights obtained from the analytics and decision systems. The third database is the ontology-based knowledge graph, which enables shared semantics, data linkage, and both user and contextual knowledge storage. The summary of key data, knowledge, and insight databases are summarised in Table 1.

Table 1 The summary of identified reading, knowledge, and insight dataset within the proposed DT

Reading database	Knowledge database	Insight database
Flight and environmental parameters		
Air humidity (g/m^3) Aircraft altitude (m) Aircraft cruising altitude (m) Aircraft airspeed (mph) Aircraft ground speed (mph) Engine inlet air temperature (°C) Engine revolutions/minute (RPM) Engine running temperature (°C) Engine coolant temperature (°C) Engine oil temperature (°C) HPT rotation speed (RPM) Intake-air temperature (°C) …	Air density (kg/m^3) Air humidity effect (%) Aircraft speed limits (mph) Aircraft ceiling (ft) Aircraft required take-off distance(ft) Aircraft climb rate (ft/min) Aircraft controllability speeds (mph) Optimal engine temperature/limit (°C) RPM error (%)	Aircraft routing Environmental control Flight planning
Engine specification and performance parameters		
Compressor inlet pressure (bar) Compressor discharge pressure (bar) Engine number of cylinders Engine cylinder bore (in) Engine turbocharger (in) Rated power (hp or kW) Engine ignition angle (°) Engine ignition timing (°) Engine stroke cycle Engine vibration (m/s^2) Engine net thrust (kg) Engine oil pressure (bar) Engine inlet air pressure (bar) Engine incoming air velocity (m/s) Exhaust gas (EWG) velocity (m/s) Exhaust gas response time Gas valve actuator response time Gas/air valve response time Intake variable valve timing …	Boost pressure error (%) Engine performance trends Engine efficiency level (%) Engine power output (hp or kW) Engine load (%) Nozzle adjustment Optimal ignition timing (°) Optimal pressure range/limit (bar) RPM limits for different flight RPM trends for health analysis Thermal efficiency level (%) Thermal capacity (J/°C) Thermal stress analysis Thrust (N) Thrust variation Turbine performance Velocity limits (m/s) Warning thresholds	Anomaly detection Engine health monitoring Oil system health control Power control system

(continued)

Table 1 (continued)

Reading database	Knowledge database	Insight database
Emission and combustion parameters		
Exhaust gas (EWG) temperature Exhaust variable valve timing EWG recirculation surface tension Fuel density (kg/m^3) Fuel pressure (bar) Fuel temperature Fuel response time Fuel kinematic viscosity (mm^2/s) Greenhouse gas emission (GHG) composition Throttle valve position Bleed valve actuator Stator vane actuator Clearance control actuator Number of flying hours (h) Number of flight cycle …	Air mass flow rate of air (kg/s) Aircraft fuel efficiency (%) CO_2e emission (tonnes) Energy consumption (KWh) Energy conservation (KWh) Emission index (EI) Flight trends/profile (optimal) Fuel consumption trends Fuel mass flow rate (kg/s) Fuel manifold pressure error (%) Ignition timing effect on emissions Optimal fuel consumption (kg/kN/h) SAF blended ratio (%) Soot emissions index Waste heat recovery Active clearance control	Corrective actions to reduce emissions Energy consumption control Energy conservation control Energy wastage control Energy control system Dual fuel control

Cognition Module The cognition module is one of the vital components in the proposed DT system, and its function is to generate an optimal sequence of control signals for action based on the current system state, user inputs, and given constraints to achieve expected behaviours. Referring to the definition in [72], the cognition module is a kind of intelligent agent that can perceive its environment and current states, meanwhile taking actions to achieve goals, which is similar to a human's advanced cognition capability, and it is called agent-based method.

Figure 7 shows how the cognition module works with human beings together to operate the engine. The workflow can be divided into three subtasks listed as follows:

- *Observation*: Characterise the running state of the engine along with the situations of its surroundings.
- *Reasoning*: Derive system instructions to guide the system adjustment.
- *Action*: Obtain the controller parameters from reasoning results with user inputs to adapt the adjustment.

The first subtask can be described as a problem of developing context-aware applications, where context is defined as any piece of information used for representing the situation of an entry [73]. The second subtask can be regarded as a decision-making problem, while the final one is about control strategy acquisition. For the context-aware issue in the observation section, the main challenge is

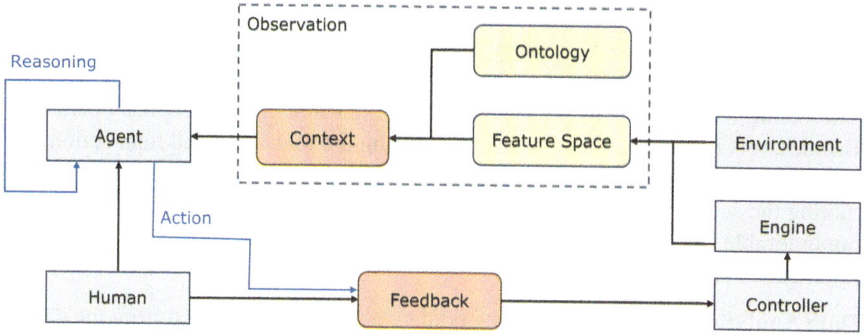

Fig. 7 The diagram of the agent-based cognition module

constructing mapping between the raw data from sensors on the engine and the domain-specific information, such as concepts of entities and related relationships between these concepts [74]. Thanks to the great leap in machine learning, nonlinear data-driven models like neural networks can be trained for various purposes to elaborate low-level sensory data to abstract feature representation (feature space) [75], enabling the system to perform some advanced tasks such as data clustering, abnormal detection, and fault prediction. Nevertheless, the latent feature space learned by machines, for instance, a high-dimensional vector space [76], cannot be treated as context information modelling because of the lack of common sense and knowledge from human beings, which is difficult to capture from pure data. To address this problem, ontology-based context modelling is proposed to explicitly represent entries and relations among them in a machine-understandable way [77], such as a graph structure format, allowing the machine to do more advanced semantic reasoning [78]. A mixture of ontology-based context modelling and data-driven analysis approach shed light on building a context-aware system, while its implementation and application in the aero industry are still under investigation.

As for the decision-making in the reasoning section, the machine is expected to select an optimal action to step toward the final goal from a provided solution base or even derive one by itself from scratch. For instance, in the aircraft controller diagram, for the sustainable objective, fuel flow is the most critical variable that needs to be optimised in the decision-making process. Therefore, in this case, the decision-making problem can be further narrowed down to how to derive the optimal fuel flow and relevant parameters according to the current context and requirements of human beings. Then, the system can either select a fuel flow according to a hardcore, well-defined rule-base system or it can learn to obtain the rate by learning. From this perspective, the reasoning is coupled with the action section, in which the decision-making is a reasoning process to determine the variant of the fuel flow in a quantitative way, indicating the trend of adjustment (increase or decrease), while the final controller parameters computation is to drive the real system with accurate signals. In the review [32], several promising methods for synthesising reasoning and action have been introduced, such as reinforcement learning, iterative learning, and event-based learning. By these methods, the agent can learn how to choose a set of parameters based on the environment without hard programming. The result will be sent to the controller as feedback for the engine operations. Besides, it is important to emphasise that during the reasoning and final action process, human-in-the-loop is still required for monitoring, validating, and finally promoting the action to ensure operational safety and reliability. The human factor is a nonignorable consideration for designing a cognition module based on artificial intelligence [79].

Data Analytics The data analytics module uses the insight gained from the data to determine the performance of the virtual representation of the aircraft engine throughout its lifecycle. The module determines how the aircraft engine can operate efficiently in its current state. Three main modules of analytics are presented in the architecture. The first module, namely AI-Machine Learning, consists of the application of data-driven approaches to achieve the intended DT goal. The second module is the Expert system, where the physical knowledge of the system can be employed. The last module, which is the simulation, can enable future health management prognosis and 'what if' type of scenario optimisation of the DT. These modules enable offline learning (model training) as well as offline and near-time analytics. The learned model can then be employed in the cognition module for real-time analysis.

During flight operations, the carbon emission from an aircraft engine cannot be directly measured. It can only be inferred from measured or controlled variables [80]. The fuel metering unit in the fuel control system of an aircraft supplies fuel through the fuel nozzle to the combustor based on the pilot's thrust demand, including any safety limits, and on the mass of compressed air flowing through the combustor [81]. This allows the assumption that the combustion of jet fuel is a stoichiometric chemical reaction. Consequently, the mass of CO_2 emitted can be directly linked to the amount of fuel burnt based on the chemical composition of the fuel used [80]. Hence, the carbon emission from an aircraft engine during flight operations can be calculated using fuel use data. The approach recommended by the ICAO (International Civil Aviation Organisation) and adopted by airline operators in the aerospace industry aligns with the monitoring, reporting, and verification requirements of the CORSIA (Carbon Offsetting and Reduction Scheme for International Aviation), which is a voluntary carbon offset and emission reduction scheme that has been applied to international aviation since 2019. The approach suggests the use of an emission factor and quantified data related to the activity associated with the release of the pollutant. The activity data considered is the amount of fuel consumed per flying hour over a forecasted period. The mapping of this activity data into carbon emissions is performed using a fuel conversion factor (FCF). The generic formula for carbon emission quantification during flight operations is shown in Eq. 1.

$$\text{Emissions}\left[\text{kg}\,\text{CO}_2\text{e}\right] = \text{Fuel flowrate}\left[\frac{\text{kg}}{h}\right] \times \text{Flying hours}\left[h\right] \times \text{FCF}\left[\frac{\text{kg}\,\text{CO}_2\text{e}}{\text{kg}}\right] \tag{1}$$

Currently, the fuel flow rate used is a fleet-averaged value over a forecasted period for each engine type over the take-off, cruising, and landing phases. As for the FCF, it is a constant parameter that depends on the type of fuel used, and what part of the value chain is being considered by the stakeholder—Tank-to-Wake (TTW) or Well-to-Wake (WTW) [75, 82]. For example, consider the emissions resulting from the combustion of Jet-A fuel during flight operations; using the ICAO guidelines, the TTW FCF is 3.16 kgCO_2/kg fuel while WTW FCF is about 3.88 kg CO_2e/kg fuel. TTW FCF calculates the amount of CO_2 emitted downstream based on the stoichiometric chemical reaction of 1 kg of jet fuel. On the other hand, WTW FCF accounts for both the upstream and downstream greenhouse gas emissions based on the life cycle emission value of jet fuel. The engine control system works with the fuel control system to determine and supply the required fuel flow rate to the combustor. Consequently, the decision-making efficiency and effectiveness of these systems have a direct impact on carbon emissions.

Interface The Interface Module holds a significant role within our digital twin architecture developed for aircraft engines. By integrating Augmented Reality/Virtual Reality (AR/VR), text-based interactions, and a dashboard, this module facilitates human-machine interaction, connecting the digital twin system with human users such as engineers, maintenance workers, and operators. To enable seamless interaction between these two parties, the module is designed to fulfil two primary functions. Firstly, it visually communicates real-time operating data from the engine system and other modules within the digital twin, providing users with a clear understanding of the current situation. Secondly, it translates user commands into machine language, thereby controlling the digital twin's operations, and subsequently feeds back operation outcomes to the physical engine system. This interface module delineates how human users engage within this digital twin system. Based on the power of AR/VR, the Interface Module offers an immersive experience, enabling engineers and operators to engage with virtual replicas of the engine components, visualise intricate processes, and conduct real-time assessments; this interaction facilitates maintenance simulations and diagnostic evaluations.

The Text component of the Interface Module establishes a bridge between human and machine data, engineers can communicate seamlessly with the engine's digital twin, extracting critical insights, querying performance parameters, and receiving a concise yet comprehensive response with a user-friendly interface. This communication empowers engineers to quickly understand the operation of the engine and respond promptly to complex problems, thereby reducing operational downtime. The dashboard provides a consolidated overview of the engine's operational status, performance metrics, and predictive analytics as a supplement to these immersive textual elements. This real-time visualisation empowers operators and engineers with actionable information, facilitating resource allocation and adaptive maintenance practices.

Cyber Security and Resilience The cross-system security platform/resilience DT component is a module that communicates with all the modules in the virtual space. In the digital twin system, the data collected in the physical space is transferred to the virtual space for a series of management and analysis, and then feedback to the physical space; the entire process is like an ever-refined circle. During this process, the communication between all modules involves security issues. Therefore, this module's existence ensures that the system operates smoothly in a safe state. Similarly, the resilience proposed in this architecture refers to the fact that the digital twin, its own system, is resilient. In the digital twin system, the virtual space is a real-time representation of the physical space, but the physical space and the virtual space cannot be kept 100% the same; with the change of time, with the update of data iteration, this part of the design is to ensure that the virtual space in a specific range of deviation from the physical space is to form an accurate description.

4.2.3 Edge Layer

In the proposed DT architecture, the 'Edge Layer' provides the infrastructure that links the physical space to the virtual space and an edge computing module that allows Real-time analytic with offline build analytics in the data analytic layer. The edge layer allows data collection and data transmission from physical to virtual space, and real-time feedback from processed data within the virtual space to the physical space. Data collection occurs through IoT devices, sensors, cameras, and other data sources in the physical space about the engine and relevant processes, which are monitors for energy efficiency, and transmitted to the 'IoT devise management' module. In some cases, some data measures are directly collected and locally analysed by the IoT devices at the edge layer (edge computing module) and then transmitted to the 'feedback' module within the virtual space. Finally, the edge layer also enables real-time feedback from the 'cognition module' to the physical space using the feedback module and several actuators. This automated feedback allows some of the decision-making and actions within the engine without relying on the 'interface' module and interactions from 'end-users'.

5 Discussion

This study highlights that integrating Digital Twin (DT) technology with aircraft engines is a pivotal advancement for the sustainable aviation industry. This research work contributed to the body of knowledge by presenting a DT architecture designed to address the challenge of data availability and inadequate data quality in aircraft engine control systems, with the aim of enhancing engine sustainability and efficiency. The importance of DT applications for sustainable systems and energy efficiency is highlighted in several studies [11, 13, 38]. However, the implementation of DT for energy and fuel efficiency in aircraft engines is limited and has not been discussed thoroughly in the literature. Nevertheless, the impact of DT on energy consumption and CO_2e footprint reduction in complex engineering assets is evident [13]. Moreover, the capability of DT regarding real-time monitoring, proactive anomaly detection, forecasting, and automating feedback allows intelligent and effective decision-making within automated energy control systems, in which the literature refers to such DT as a solution for Energy IoT (EIoT) [20]. The integration of DT technology with the energy management systems and as part of the decision support systems can improve the energy distribution, optimisation and, therefore, performance of the asset [23, 44].

This study considered the Ardebili et al. [20] observation and presented a DT-enabled control system for the thrust control of an aircraft engine. This study focuses on the engine control system to demonstrate the DT integration and its application for delivering sustainable aviation. A typical control system, including the fuel system, is demonstrated in Fig. 5. The integrated system determines and supplies the required fuel flow rate to the combustor. Consequently, the decision-making efficiency and effectiveness of these systems and their ability to account for changes in operation and environmental conditions have a direct impact on the sustainability indicator. The fuel flow rate is the critical indicator to assess CO_2e emissions and sustainability during the flight operation (see Eq. 1). In aircraft engines, the fuel flow rate to achieve a desired engine pressure ratio (EPR) or fan speed (N1) is not only dependent on the pilot command, and operational and safety limits but also on the changes in operational and environmental conditions of the engine. During flight operations, the components of the aircraft engine degrade. This can take the form of hardware ageing deterioration and contamination by airborne particles. The result is an increased Specific Fuel Consumption (SFC) to achieve the desired thrust [68, 69]. The proposed DT for the control system makes it self-aware of the engine's current state and gives the control system the ability to adapt to changes in operational and environmental conditions dynamically with a view to optimising energy efficiency and fuel consumption. As a result, the aircraft engine does not operate in a conservative manner; it will have a sustainable and efficient operation based on the extended capability of the DT-enabled control system, which allows continuous real-time collection, assessment, analysis, and prediction of fuel and energy consumption.

The integration of DT principles in the optimisation of engine sustainability-related parameters aligns with the aviation industry's drive for improved energy efficiency and environmentally conscious propulsion systems. The literature highlights the effectiveness of the DT-driven optimisation approach in refining operations [36, 41, 44, 45]. By incorporating real-time data into the DT model and integrating feedback into key performance indicators, this approach facilitates iterative optimisation adjustments and enhances energy efficiency strategies. These studies collectively emphasise DT's potential to refine engine operation, conserve energy, and optimise system efficiency. In addition, the DT system enables real-time data using embedded IoT devices and empowers precise monitoring and analysis of engine parameters and provides a comprehensive sustainability assessment of engine behaviour across diverse conditions. This data-driven approach facilitates the refinement of parameters for optimal performance. Additionally, the dynamic feedback loop between physical and virtual spaces, facilitated by DT, ensures continuous parameter monitoring and adjustment. This equips engineers to not only enhance real-time performance but also adapt to evolving operational demands. As a result, DT goes beyond traditional static models by introducing dynamic, data-informed approaches, offering significant improvements in aircraft engine operational efficiency, reliability, and sustainability.

Recent studies highlight a growing interest in integrating semantic technologies with DTs, particularly the association of ontology and DTs. Boschert et al. [83] referred to the next-gen Digital Twin, known as nexus, as one that fosters system interconnectivity through semantic tools, such as ontologies. Akroyd et al. [83] proposed the idea of the Universal Digital Twin, a DT that employs a dynamic knowledge graph to enhance cross-domain interoperability. Lu et al. [83] introduced the idea of the Cognitive Twin (CT), which is a DT fortified with enriched semantic abilities to aid IoT systems or decision-making processes in manufacturing [84]. Li et al. [85] applied the cognitive twin framework to facilitate the co-simulation of intricate engineering assets, integrating multiple digital tools. Abburu et al. [86] developed the COGNITWIN software toolkit tailored for the process industry, viewing the CT as an advanced form of the Hybrid DT (HT). The latter was seen as an advancement of the conventional DT. Combining ontology concepts with digital twins improves interoperability and integration with physical systems and equipment [87]. The emergence of AI technology has significantly influenced the evolution of DT technologies primarily due to two pivotal factors. Firstly, its ability to extract insightful information from vast and intricate non-linear datasets enables an efficient recognition and analysis of system states. Secondly, the integration of decision-making processes through algorithms like fuzzy logic, bio-inspired algorithms, and reinforcement learning allows for automated decision support, which can be coupled with the controller for system operation [49]. Although intelligent engine with relevant digital technology such as IoT, artificial intelligence, and knowledge engineering is now an active research topic, the business benefits should also be simultaneously considered when introducing the above technologies into the current engine lifecycle to avoid a technique-sage development [74], which is a less investigated topic.

6 Conclusion

In the aviation industry, there are several approaches to improve sustainability, such as flight efficiency improvement, sustainable aviation fuel (SAF), renewable and mixed energies, and decarbonisation, as illustrated in Fig. 1. This paper focuses on the application of some emerging technologies such as digital twin (DT), artificial intelligence (AI), and ontologies, to enhance fuel and energy efficiency within an aircraft engine. A thorough systematic literature review (see Fig. 4) is conducted to identify the research contribution, gaps, and data specification for this research. A DT architecture for a sustainable control system of an aircraft is proposed in this paper (see Fig. 6). The typical functional flow diagram of an engine control system is presented in Fig. 5. The DT architecture is developed based on a thorough literature review around sustainability and digital twin, and elicitation of expert knowledge using the Delphi method. The architecture follows the existing guidelines and standards in ISO 23247-1:2021. It is argued that the proposed DT architecture enhances energy efficiency, fuel consumption, and environmentally sustainable operation of an aircraft. The DT architecture enables (near) real-time data collection, monitoring, and analysis from the physical space—including aircraft engine and operator—and optimises and feeds back the insights to the end users, including designers, engineers, operators, and maintainers. The DT system is composed of physical aircraft, end users, IoT devices, actuators, and the edge layer that integrates with the virtual space, including IoT devise management, resource access and interchange, data analytics, cognition module, interface, and the cross-system securing platform. The proposed DT architecture considered several modules within the virtual space to enable intelligent data management, analysis, and feedback with a view to enhancing energy efficiency, fuel efficiency, and environmental sustainability. The resource access and interchange module proposed an integrated API that connects with an insight and decision database, a reading database, and an ontology-based knowledge base. The list of key databases is summarised in Table 1.

Further studies should explore the ontology technologies for the DT knowledge base development and implementation. Ontologies are formal representations of knowledge that define concepts and relationships within a domain. Combining these two concepts can offer several benefits in creating and managing Digital Twins, such as semantic interoperability, data integration, modelling complex systems, knowledge sharing and reuse, reasoning and inference, adaptability, lifecycle management, and collaboration. To implement an ontology-based approach for developing a DT technology, you would typically define the ontology that describes the concepts, properties, and relationships relevant to the physical entity or system. This ontology would serve as the basis for creating digital models, simulations, and data integration processes. Different tools and technologies exist for building and managing ontologies, such as the Web Ontology Language (OWL) and ontology development platforms.

References

1. Burrus, D. L. (1989). Application of numerical models for predictions of turbine engine combustor performance. In *Proceedings of the ASME turbo expo*. American Society of Mechanical Engineers (ASME). https://doi.org/10.1115/89GT251

2. Mongia, H. C. (2010). On continuous NOx reduction of aero-propulsion engines. In *48th AIAA aerospace sciences meeting including the new horizons forum and aerospace exposition*. https://doi.org/10.2514/6.2010-1329
3. Innocenti, A., Andreini, A., Facchini, B., & Peschiulli, A. (2017). Numerical analysis of the dynamic flame response of a spray flame for aero-engine applications. *International Journal of Spray and Combustion Dynamics, 9*(4), 310–329. https://doi.org/10.1177/1756827717703577
4. Von Der Bank, R., Donnerhack, S., Rae, A., Cazalens, M., Lundbladh, A., & Dietz, M. (2014). LEMCOTEC—Improving the core-engine thermal efficiency. In *Proceedings of the ASME turbo expo*. American Society of Mechanical Engineers (ASME). https://doi.org/10.1115/GT2014-25040
5. Liu, X., Zhao, D., Guan, D., Becker, S., Sun, D., & Sun, X. (2022). Development and progress in aeroacoustic noise reduction on turbofan aeroengines. *Progress in Aerospace Sciences, 130*. https://doi.org/10.1016/j.paerosci.2021.100796
6. Otto, M., Vesely, L., Kapat, J., Stoia, M., Applegate, N. D., & Natsui, G. (2022). Ammonia as an aircraft fuel: Thermal assessment from airport to wake. In *Proceedings of the ASME turbo expo*. American Society of Mechanical Engineers (ASME). https://doi.org/10.1115/GT2022-84359
7. Energy Technology Perspectives. (2020). [Online]. Available: https://www.iea.org/reports/energy-%0ATechnology-perspectives-2020.
8. Coelho Barbosa, F. (2021). Aircraft aerodynamic technology review—A tool for aviation performance and sustainability improvement. In *SAE technical papers*. SAE International. https://doi.org/10.4271/2022-36-0022
9. Jamwal, A., Agrawal, R., Sharma, M., & Giallanza, A. (2021). Industry 4.0 technologies for manufacturing sustainability: A systematic review and future research directions. *Applied Sciences (Switzerland), 11*(12). https://doi.org/10.3390/app11125725
10. Davis, S. J., et al. (2018). Net-zero emissions energy systems. *Science, 360*(6396), eaas9793. https://doi.org/10.1126/science.aas9793
11. Abdoune, F., Ragazzini, L., Nouiri, M., Negri, E., & Cardin, O. (2023). Toward digital twin for sustainable manufacturing: A data-driven approach for energy consumption behavior model generation. *Computers in Industry, 150*. https://doi.org/10.1016/j.compind.2023.103949
12. Yan, Q., Wang, B., & Ye, Z. (2021). Multi-physical coupled simulation on fuel cooling Shell of electric fuel pump. In *ASME international mechanical engineering congress and exposition, proceedings (IMECE)*. American Society of Mechanical Engineers (ASME). https://doi.org/10.1115/IMECE2021-73190
13. Zhang, M., Zuo, Y., & Tao, F. (2018). Equipment energy consumption management in digital twin shop-floor: A framework and potential applications. In *ICNSC 2018—15th IEEE international conference on networking, sensing and control* (pp. 1–5). Institute of Electrical and Electronics Engineers. https://doi.org/10.1109/ICNSC.2018.8361272
14. Raman, V., & Hassanaly, M. (2019). Emerging trends in numerical simulations of combustion systems. *Proceedings of the Combustion Institute, 37*(2), 2073–2089. https://doi.org/10.1016/j.proci.2018.07.121
15. Taluru, D., & Allabanda, R. P. U. (2019). Application of data analytics in gas turbine engines. In *ASME 2019 gas turbine India conference, GTINDIA 2019*. American Society of Mechanical Engineers (ASME). https://doi.org/10.1115/GTINDIA2019-2557
16. Kychkin, A., Deryabin, A., Vikentyeva, O., & Shestakova, L. (2019). Architecture of compressor equipment monitoring and control cyber-physical system based on influxdata platform. In *2019 international conference on industrial engineering, applications and manufacturing, ICIEAM 2019*. Institute of Electrical and Electronics Engineers. https://doi.org/10.1109/ICIEAM.2019.8742963
17. Saad, A., Faddel, S., & Mohammed, O. (2020). IoT-based digital twin for energy cyber-physical systems: Design and implementation. *Energies (Basel), 13*(18). https://doi.org/10.3390/en13184762

18. Grieves, M. (2014). Digital twin: Manufacturing excellence through virtual factory replication. *White Paper, 1*, 1–7.
19. D'Amico, R. D., Erkoyuncu, J. A., Addepalli, S., & Penver, S. (2022). Cognitive digital twin: An approach to improve the maintenance management. *CIRP Journal of Manufacturing Science and Technology, 38*, 613–630. https://doi.org/10.1016/J.CIRPJ.2022.06.004
20. Ardebili, A. A., Longo, A., & Ficarella, A. (2021). Digital twin (DT) in smart energy systems—Systematic literature review of DT as a growing solution for energy internet of the things (EIoT). In *E3S web of conferences*. EDP Sciences. https://doi.org/10.1051/e3sconf/202131209002
21. Contini, G., Peruzzini, M., Bulgarelli, S., & Bosi, G. (2023). Developing key performance indicators for monitoring sustainability in the ceramic industry: The role of digitalization and industry 4.0 technologies. *Journal of Cleaner Production, 414*. https://doi.org/10.1016/j.jclepro.2023.137664
22. Gao, Y., Chang, D., & Chen, C.-H. (2023). A digital twin-based approach for optimizing operation energy consumption at automated container terminals. *Journal of Cleaner Production, 385*. https://doi.org/10.1016/j.jclepro.2022.135782
23. Li, Y., Wang, S., Duan, X., Liu, S., Liu, J., & Hu, S. (2021). Multi-objective energy management for Atkinson cycle engine and series hybrid electric vehicle based on evolutionary NSGA-II algorithm using digital twins. *Energy Conversion and Management, 230*. https://doi.org/10.1016/j.enconman.2020.113788
24. Yu, W., Patros, P., Young, B., Klinac, E., & Walmsley, T. G. (2022). Energy digital twin technology for industrial energy management: Classification, challenges and future. *Renewable and Sustainable Energy Reviews, 161*. https://doi.org/10.1016/j.rser.2022.112407
25. Davies, O., Makkattil, A., Jiang, C., & Farsi, M. (2022). A digital twin design for maintenance optimization. *Procedia CIRP, 109*, 395–400. https://doi.org/10.1016/J.PROCIR.2022.05.268
26. Badakhshan, E., & Ball, P. (2021). Reviewing the application of data driven digital twins in manufacturing systems: A business and management perspective. In A. Dolgui, A. Bernard, D. Lemoine, C. G. Von, & D. Romero (Eds.), *IFIP advances in information and communication technology* (pp. 256–265). Springer Science and Business Media Deutschland GmbH. https://doi.org/10.1007/978-3-030-85910-7_27
27. Franciosi, C., Miranda, S., Veneroso, C. R., & Riemma, S. (2022). Improving industrial sustainability by the use of digital twin models in maintenance and production activities. In B. Barbieri, D. Romero, C. Emmanouilidis, A. Parlikad, & S. Sepideh (Eds.), *IFAC-PapersOnLine* (pp. 37–42). Elsevier B.V. https://doi.org/10.1016/j.ifacol.2022.09.215
28. Alford, M., Udugama, I., Yu, W., & Young, B. (2022). Flexible digital twins from commercial off-the-shelf software solutions: A driver for energy efficiency and decarbonisation in process industries? *Chemical Product and Process Modeling, 17*(4), 395–407. https://doi.org/10.1515/cppm-2021-0045
29. Pires, F., Ahmad, B., Moreira, A. P., & Leitao, P. (2021). Digital twin based what-if simulation for energy management. In *Proceedings—2021 4th IEEE international conference on industrial cyber-physical systems, ICPS 2021* (pp. 309–314). Institute of Electrical and Electronics Engineers. https://doi.org/10.1109/ICPS49255.2021.9468224
30. Tedstone, L. (2021). Age doesn't matter: Digitizing the as-is condition of brownfield assets is part of a leaner & greener future. In *Society of petroleum engineers—Abu Dhabi international petroleum exhibition and conference, ADIP 2021*. Society of Petroleum Engineers. https://doi.org/10.2118/208218-MS
31. Ezhilarasu, C. M., Skaf, Z., & Jennions, I. K. (2021). A generalised methodology for the diagnosis of aircraft systems. *IEEE Access, 9*, 11437–11454. https://doi.org/10.1109/ACCESS.2021.3050877
32. Diaz, R. A. C., Ghita, M., Copot, D., Birs, I. R., Muresan, C., & Ionescu, C. (2020). Context aware control systems: An engineering applications perspective. *IEEE Access, 8*, 215550–215569. https://doi.org/10.1109/ACCESS.2020.3041357
33. Rucco, C., Longo, A., & Zappatore, M. (2022). Supporting energy digital twins with cloud data spaces: An architectural proposal. In P. L. Mazzeo, C. Distante, E. Frontoni, & S. Sclaroff

(Eds.), *Lecture notes in computer science (including subseries lecture notes in artificial intelligence and lecture notes in bioinformatics)* (pp. 47–58). Springer Science and Business Media Deutschland GmbH. https://doi.org/10.1007/978-3-031-13324-4_5
34. Deena, G., Gulati, K., Jha, R., Bajjuri, U. R., Sahithullah, M., & Singh, M. (2022). Artificial intelligence and a digital twin are effecting building energy management. In *Proceedings of the 2022 international conference on innovative computing, intelligent communication and smart electrical systems, ICSES 2022*. Institute of Electrical and Electronics Engineers. https://doi.org/10.1109/ICSES55317.2022.9914233
35. Yang, X., Maiti, A., Jiang, J., & Kist, A. (2022). Forecasting and monitoring smart buildings with the internet of things, digital twins and blockchain. In M. E. Auer, K. R. Bhimavaram, & X. Yue (Eds.), *Lecture notes in networks and systems* (pp. 213–224). Springer Science and Business Media Deutschland GmbH. https://doi.org/10.1007/978-3-030-82529-4_21
36. Bartolucci, L., Cennamo, E., Cordiner, S., Mulone, V., Pasqualini, F., & Aimo Boot, M. (2022). Fuel cell hybrid electric vehicle: Driving cycle impact on control strategy design and system performances. In *SAE technical papers*. SAE International. https://doi.org/10.4271/2022-24-0003
37. Ioshchikhes, B., Borst, F., & Weigold, M. (2022). Assessing energy efficiency measures for hydraulic systems using a digital twin. In A. Valente, E. Carpanzano, & C. Boer (Eds.), *Procedia CIRP* (pp. 1232–1237). Elsevier B.V. https://doi.org/10.1016/j.procir.2022.05.137
38. Calabuig-Moreno, R., Temes-Cordovez, R., & Orozco-Messana, J. (2022). Neighbourhood digital modelling of energy consumption for carbon footprint assessment. In J. R. Littlewood, R. J. Howlett, & L. C. Jain (Eds.), *Smart innovation, systems and technologies* (pp. 541–551). Springer Science and Business Media Deutschland GmbH. https://doi.org/10.1007/978-981-16-6269-0_45
39. Adu-Kankam, K. O., & Camarinha-Matos, L. M. (2023). Modeling collaborative behaviors in energy ecosystems. *Computers, 12*(2), 39. https://doi.org/10.3390/computers12020039
40. Barykin, S. E., Sergeev, S. M., Provotorov, V. V., Lavskaya, K. K., Kharlamov, A. V., & Kharlamova, T. L. (2022). Energy efficient digital omnichannel marketing based on a multidimensional approach to network interaction. *Frontiers in Energy Research, 10*. https://doi.org/10.3389/fenrg.2022.946588
41. Assad, F., Konstantinov, S., Ahmad, M. H., Rushforth, E. J., & Harrison, R. (2021). Utilising web-based digital twin to promote assembly line sustainability. In *Proceedings—2021 4th IEEE international conference on industrial cyber-physical systems, ICPS 2021* (pp. 381–386). Institute of Electrical and Electronics Engineers. https://doi.org/10.1109/ICPS49255.2021.9468209
42. Zhao, Q., Chen, S., Wang, X., Tian, J., Zhao, R., & Yang, J. (2022). Research on key technology of digital twin and its application in integrated energy system. In *2022 12th international conference on power and energy systems, ICPES 2022* (pp. 950–956). Institute of Electrical and Electronics Engineers. https://doi.org/10.1109/ICPES56491.2022.10073431
43. Sun, L., Liu, T., Wang, D., Huang, C., & Xie, Y. (2022). Deep learning method based on graph neural network for performance prediction of supercritical CO2 power systems. *Applied Energy, 324*. https://doi.org/10.1016/j.apenergy.2022.119739
44. Xu, Z., et al. (2021). Digital twin-driven optimization of gas exchange system of 2-stroke heavy fuel aircraft engine. *Journal of Manufacturing Systems, 58*, 132–145. https://doi.org/10.1016/j.jmsy.2020.08.002
45. Wang, J. F., Huang, Y. Q., & Tang, D. L. (2021). A digital twin simulator for real time energy saving control of serial manufacturing system. In *2021 IEEE international conference on real-time computing and robotics, RCAR 2021* (pp. 720–725). Institute of Electrical and Electronics Engineers. https://doi.org/10.1109/RCAR52367.2021.9517579
46. Li, H., et al. (2022). Data-driven hybrid petri-net based energy consumption behaviour modelling for digital twin of energy-efficient manufacturing system. *Energy, 239*. https://doi.org/10.1016/j.energy.2021.122178

47. Krommes, S., & Tomaschko, F. (2023). Conceptual framework of a digital twin fostering sustainable manufacturing in a brownfield approach of small volume production for SMEs. In H. Kohl, G. Seliger, & F. Dietrich (Eds.), *Lecture notes in mechanical engineering* (pp. 519–527). Springer Science and Business Media Deutschland GmbH. https://doi.org/10.1007/978-3-031-28839-5_58
48. Dongyun, D., Zheng, W., Yimin, Y., Zhongqing, S., Huisheng, Y., & Weiyun, J. (2022). Research on intelligent online operation and maintenance system of 3D visualization hydrogen production and energy storage Power Station. In *2022 4th international conference on smart power and internet energy systems, SPIES 2022* (pp. 2128–2133). Institute of Electrical and Electronics Engineers. https://doi.org/10.1109/SPIES55999.2022.10081961
49. Zhang, Z., Zeng, Y., Liu, H., Zhao, C., Wang, F., & Chen, Y. (2022). Smart DC: An AI and digital twin-based energy-saving solution for data centers. In P. Varga, L. Z. Granville, A. Galis, I. Godor, N. Limam, P. Chemouil, J. Francois, & M.-O. Phai (Eds.), *Proceedings of the IEEE/IFIP network operations and management symposium 2022: Network and service management in the era of cloudification, softwarization and artificial intelligence, NOMS 2022*. Institute of Electrical and Electronics Engineers Inc. https://doi.org/10.1109/NOMS54207.2022.9789853
50. Passath, T., Huber, C., Kohl, L., Biedermann, H., & Ansari, F. (2021). A knowledge-based digital lifecycle-oriented asset optimisation. In *Tehnicki Glasnik* (pp. 226–234). University North. https://doi.org/10.31803/tg-20210504111400
51. Deon, B., et al. (2022). Digital twin and machine learning for decision support in thermal power plant with combustion engines. *Knowledge-Based Systems, 253*. https://doi.org/10.1016/j.knosys.2022.109578
52. Gebauer, M., Blejchař, T., Brzobohatý, T., Karásek, T., & Nevřela, M. (2022). Determination of aerodynamic losses of electric motors. *Symmetry (Basel), 14*(11). https://doi.org/10.3390/sym14112399
53. Pourfarzaneh, H., Hajilouy-Benisi, A., & Farshchi, M. (2010). An analytical model of a gas turbine components performance and its experimental validation. In *Proceedings of the ASME turbo expo* (pp. 335–340). https://doi.org/10.1115/GT2010-23369
54. Li, Y. G., & Pilidis, P. (2010). GA-based design-point performance adaptation and its comparison with ICM-based approach. *Applied Energy, 87*(1), 340–348. https://doi.org/10.1016/j.apenergy.2009.05.034
55. Xin, X., Tan, J., Liu, Z., Sui, Y., & Ding, J. (2022). Research progress on forward design of gas turbine. *Jixie Gongcheng Xuebao/Journal of Mechanical Engineering, 58*(17), 191–205. https://doi.org/10.3901/JME.2022.17.191
56. Sciatti, F., Tamburrano, P., De Palma, P., Distaso, E., & Amirante, R. (2022). Detailed simulations of an aircraft fuel system by means of Simulink. In *Journal of physics: Conference series*. Institute of Physics. https://doi.org/10.1088/1742-6596/2385/1/012033
57. Stoumpos, S., Bolbot, V., Theotokatos, G., & Boulougouris, E. (2022). Safety performance assessment of a marine dual fuel engine by integrating failure mode, effects and criticality analysis with simulation tools. *Proceedings of the Institution of Mechanical Engineers Part M: Journal of Engineering for the Maritime Environment, 236*(2), 376–393. https://doi.org/10.1177/14750902211043423
58. Junckes, R. F., Varnier, C. A. C., Nakirimoto, E. K., & Tavares, L. H. S. (2022). Digital twin application in thermal system with a heat source unknown. In *2022 international conference on electrical machines, ICEM 2022* (pp. 1791–1795). Institute of Electrical and Electronics Engineers. https://doi.org/10.1109/ICEM51905.2022.9910648
59. Duarte, E. P., Viegas, E. K., & Santin, A. O. (2022). A machine learning-based digital twin model for pressure prediction in the fuel injection system. In *IECON proceedings (industrial electronics conference)*. IEEE Computer Society. https://doi.org/10.1109/IECON49645.2022.9968945
60. Zhu, M., Yang, B., & Peng, C. (2022). A model-data combined driven vibration digital twin model for magnetically suspended motor. In *2022 international conference on electrical machines and systems, ICEMS 2022*. Institute of Electrical and Electronics Engineers Inc. https://doi.org/10.1109/ICEMS56177.2022.9983014

61. Dai, Y., Zhang, K., Maharjan, S., & Zhang, Y. (2021). Deep reinforcement learning for stochastic computation offloading in digital twin networks. *IEEE Transactions on Industrial Informatics, 17*(7), 4968–4977. https://doi.org/10.1109/TII.2020.3016320
62. Aliramezani, M., Koch, C. R., & Shahbakhti, M. (2022). Modeling, diagnostics, optimization, and control of internal combustion engines via modern machine learning techniques: A review and future directions. *Progress in Energy and Combustion Science, 88*, 100967. https://doi.org/10.1016/J.PECS.2021.100967
63. Ghenai, C., Husein, L. A., Al Nahlawi, M., Hamid, A. K., & Bettayeb, M. (2022). Recent trends of digital twin technologies in the energy sector: A comprehensive review. *Sustainable Energy Technologies and Assessments, 54*. https://doi.org/10.1016/j.seta.2022.102837
64. Mouzakitis, S., et al. (2023). Enabling maritime digitalization by extreme-scale analytics, AI and digital twins: The Vesselai architecture. In K. Arai (Ed.), *Lecture notes in networks and systems* (pp. 246–256). Springer Science and Business Media Deutschland GmbH. https://doi.org/10.1007/978-3-031-16075-2_16
65. Agouzzal, K., & Abbou, A. (2022). A hybrid method integrating industry 4.0's energy digitization. *WSEAS Transactions on Systems, 21*, 157–167. https://doi.org/10.37394/23202.2022.21.17
66. Lu, Y., Huang, X., Zhang, K., Maharjan, S., & Zhang, Y. (2021). Communication-efficient federated learning for digital twin edge networks in industrial IoT. *IEEE Transactions on Industrial Informatics, 17*(8), 5709–5718. https://doi.org/10.1109/TII.2020.3010798
67. Bortolini, R., Rodrigues, R., Alavi, H., Vecchia, L. F. D., & Forcada, N. (2022). Digital twins' applications for building energy efficiency: A review. *Energies (Basel), 15*(19). https://doi.org/10.3390/en15197002
68. Chen, D., & Sun, J. (2018). Fuel and emission reduction assessment for civil aircraft engine fleet on-wing washing. *Transportation Research Part D: Transport and Environment, 65*, 324–331. https://doi.org/10.1016/j.trd.2018.05.013
69. Sogut, M. Z., Yalcin, E., & Karakoc, T. H. (2017). Assessment of degradation effects for an aircraft engine considering exergy analysis. *Energy, 140*, 1417–1426. https://doi.org/10.1016/j.energy.2017.03.093
70. Csank, J., May, R., Litt, J., & Guo, T. H. (2010, October). Control design for a generic commercial aircraft engine. In *46th AIAA/ASME/SAE/ASEE joint propulsion conference & exhibit* (p. 6629).
71. Imani, A., & Montazeri-Gh, M. (2019). A min-max selector controller for turbofan engines with improvement of limit management and low computational burden. *Transactions of the Institute of Measurement and Control, 41*(1), 36–44. https://doi.org/10.1177/0142331217752043
72. Wooldridge, M. (1999). Intelligent agents. In *Multiagent systems: A modern approach to distributed artificial intelligence* (Vol. 1, pp. 27–73).
73. Abowd, G. D., Dey, A. K., Brown, P. J., Davies, N., Smith, M., & Steggles, P. (1999). Towards a better understanding of context and context-awareness. *Lecture Notes in Computer Science, 1707*, 304–307. https://doi.org/10.1007/3-540-48157-5_29/COVER
74. Xie, Y., Weber, F., & Culley, S. (2011). Opportunities and challenges for context-aware systems in aerospace industry. *Journal of Enterprise Information Management, 24*(2), 118–125. https://doi.org/10.1108/17410391111106257/FULL/PDF
75. Bengio, Y., Courville, A., & Vincent, P. (2012). Representation learning: A review and new perspectives. *IEEE Transactions on Pattern Analysis and Machine Intelligence, 35*(8), 1798–1828. https://doi.org/10.1109/TPAMI.2013.50
76. Agostinelli, S., Cumo, F., Guidi, G., & Tomazzoli, C. (2020). The potential of digital twin model integrated with artificial intelligence systems. In Z. Leonowicz (Ed.), *Proceedings—2020 IEEE international conference on environment and electrical engineering and 2020 IEEE industrial and commercial power systems Europe, EEEIC/I and CPS Europe 2020*. Institute of Electrical and Electronics Engineers. https://doi.org/10.1109/EEEIC/ICPSEurope49358.2020.9160810
77. Gruber, T. R. (1993). A translation approach to portable ontology specifications. *Knowledge Acquisition, 5*(2), 199–220. https://doi.org/10.1006/KNAC.1993.1008

78. Aguilar, J., Jerez, M., & Rodríguez, T. (2018). CAMeOnto: Context awareness meta ontology modeling. *Applied Computing and Informatics, 14*(2), 202–213. https://doi.org/10.1016/J.ACI.2017.08.001
79. Rožanec, J. M., et al. (2022). Human-centric artificial intelligence architecture for industry 5.0 applications. *International Journal of Production Research, 2023*(20), 6847–6872. https://doi.org/10.1080/00207543.2022.2138611
80. Jardine, C. N. (2009). *Calculating the carbon dioxide emissions of flights*. Available: http://www.atmosfair.de/index.php?L=3
81. Li, D., Hang, J., Li, Y., & Dong, S. (2022). Fuel flowrate control for aeroengine and fuel thermal management for airborne system of aircraft—An overview. *Applied Sciences (Switzerland), 12*(1), 279. https://doi.org/10.3390/app12010279
82. Jing, L., et al. (2022). Understanding variability in petroleum jet fuel life cycle greenhouse gas emissions to inform aviation decarbonization. *Nature Communications, 13*(1). https://doi.org/10.1038/s41467-022-35392-1
83. Rosen, R., Fischer, J., & Boschert, S. (2019). Next generation digital twin: An ecosystem for mechatronic systems? *IFAC-PapersOnLine, 52*(15), 265–270. https://doi.org/10.1016/J.IFACOL.2019.11.685
84. Rožanec, J. M., et al. (2020). Towards actionable cognitive digital twins for manufacturing. In *2020 international workshop on semantic digital twins, SeDiT 2020*. Heraklion, Greece.
85. Li, Y., Chen, J., Hu, Z., Zhang, H., Lu, J., & Kiritsis, D. (2021). Co-simulation of complex engineered systems enabled by a cognitive twin architecture. *International Journal of Production Research*. https://doi.org/10.1080/00207543.2021.1971318
86. Abburu, S., Berre, A. J., Jacoby, M., Roman, D., Stojanovic, L., & Stojanovic, N. (2020). COGNITWIN—Hybrid and cognitive digital twins for the process industry. In *Proceedings—2020 IEEE international conference on engineering, technology and innovation, ICE/ITMC 2020*.
87. Bjørnskov, J., & Jradi, M. (2023). An ontology-based innovative energy modeling framework for scalable and adaptable building digital twins. *Energy and Buildings, 292*, 113146. https://doi.org/10.1016/j.enbuild.2023.113146

Index

T. H. Karakoc et al. (eds.), *Sustainable Materials and Manufacturing Techniques in Aviation*, Sustainable Aviation, https://doi.org/10.1007/978-3-031-62987-7

The manufacturer's authorised representative in the EU is Springer Nature Customer Service Centre GmbH, Europaplatz 3, 69115 Heidelberg, Germany. If you have any concerns regarding our products, please contact ProductSafety@springernature.com

Printed and bound by CPI Group (UK) Ltd, Croydon, CR0 4YY
01/10/2026
02225880-0012